Numerical modeling and simulation of particulate fouling on structured heat transfer surfaces using multiphase Eulerian-Lagrangian LES

Numerical modeling and simulation of particulate fouling on structured heat transfer surfaces using multiphase Eulerian-Lagrangian LES

Dissertation

zur

Erlangung des akademischen Grades

Doktor-Ingenieur (Dr.-Ing.)

der Fakultät für Maschinenbau und Schiffstechnik

der Universität Rostock

---

vorgelegt von	Robert Kasper, M.Sc.
	geboren am 05. April 1988
	in Magdeburg
aus	Rostock

Rostock, 2020

Bibliografische Information der Deutschen Nationalbibliothek
Die Deutsche Nationalbibliothek verzeichnet diese Publikation in der Deutschen Nationalbibliographie; detaillierte bibliographische Daten sind im Internet über http://dnb.d-nb.de abrufbar.
1. Aufl. - Göttingen: Cuvillier, 2021
Zugl.: Rostock, Univ., Diss., 2020

Nonnenstieg 8, 37075 Göttingen
Telefon: 0551-54724-0
Telefax: 0551-54724-21
www.cuvillier.de

1. Auflage 2021
Gedruckt auf umweltfreundlichem, säurefreiem Papier aus nachhaltiger Forstwirtschaft.

ISBN 978-3-7369-7419-7
eISBN 978-3-7369-6419-8

Gutachter:

1. Gutachter:

   Prof. Dr.-Ing. habil. Nikolai Kornev,

   Lehrstuhl für Modellierung und Simulation,

   Universität Rostock

2. Gutachter:

   Prof. Dr.-Ing. Stephan Scholl,

   Institut für Chemische und Thermische Verfahrenstechnik,

   Technische Universität Braunschweig

Datum der Einreichung: 01.09.2020

Datum der Verteidigung: 26.04.2021

Danksagung

Diese Dissertation enstand im Rahmen des durch die Deutsche Forschungsgemeinschaft (DFG) geförderten Verbundprojektes „Wirkmechanismen des Partikelfoulings auf strukturierten wärmeübertragenden Oberflächen“ und ist das Ergebnis der Unterstützung vieler Menschen, bei denen ich mich herzlich bedanken möchte.

Großer Dank gebührt meinem Doktorvater Prof. Nikolai Kornev, zum einen für die Möglichkeit zur Durchführung der Promotion am Lehrstuhl für Modellierung und Simulation (LeMoS) an der Universität Rostock, zum anderen für das stets entgegengebrachte Vertrauen sowie die fachliche Unterstützung. Prof. Stephan Scholl vom Institut für Chemische und Thermische Verfahrenstechnik (ICTV) der Technischen Universität Braunschweig möchte ich für die Übernahme des Korreferats danken. In diesem Zusammenhang gilt mein Dank auch Dr. Wolfgang Augustin für die fokussierten Diskussionen und Anregungen zum Thema „Fouling“, sowie Hannes Deponte für die reibungslose und ertragreiche Zusammenarbeit.

Ich danke allen früheren und aktuellen Kollegen am LeMoS für die angenehme Arbeitsatmosphäre, die tatkräftige Unterstützung sowie den stetigen Gedankenaustausch, möchte hier jedoch Dr. Johann Turnow hervorheben, der durch seine fachliche sowie persönliche Unterstützung einen erheblichen Anteil am Gelingen meiner Promotion hat. Dr. Matthias Walter soll aufgrund seiner Unterstützung in Programmierfragen und Fragen bezüglich Hochleistungsrechentechnik sowie vieler hilfreicher, fachlicher Anmerkungen gesondert erwähnt werden. Auch meinem Zimmerkollegen Henry Plischka möchte ich dafür danken, dass er meinen „Wahnsinn“ in der Endphase der Promotion so stoisch ertragen hat und trotzdem stets ein offenes Ohr hatte.

Darüber hinaus habe ich in der gesamten Zeit meiner Promotion viel Kraft und Rückhalt durch meine Freunde erfahren, wofür ich an dieser Stelle „Danke“ sagen möchte. Ein ganz persönlicher Dank gebührt Dr. Philipp Berg, der in mir das Interesse für die (numerische) Strömungsmechanik geweckt hat und mir zudem die Möglichkeit gegeben hat, erste Erfahrungen auf diesem unerschöpflichen Forschungsgebiet zu sammeln.

Diese Arbeit widme ich meinen Eltern Kathrin und Peter, die mich in jeglicher Weise förderten und damit das Studium und die anschließende Promotion überhaupt erst ermöglicht haben, meiner Schwester Laura, sowie ganz besonders meiner Frau Nicole, für ihren bedingungslosen Zuspruch sowie ihre grenzenlose Unterstützung während der Fertigstellung dieser Arbeit.

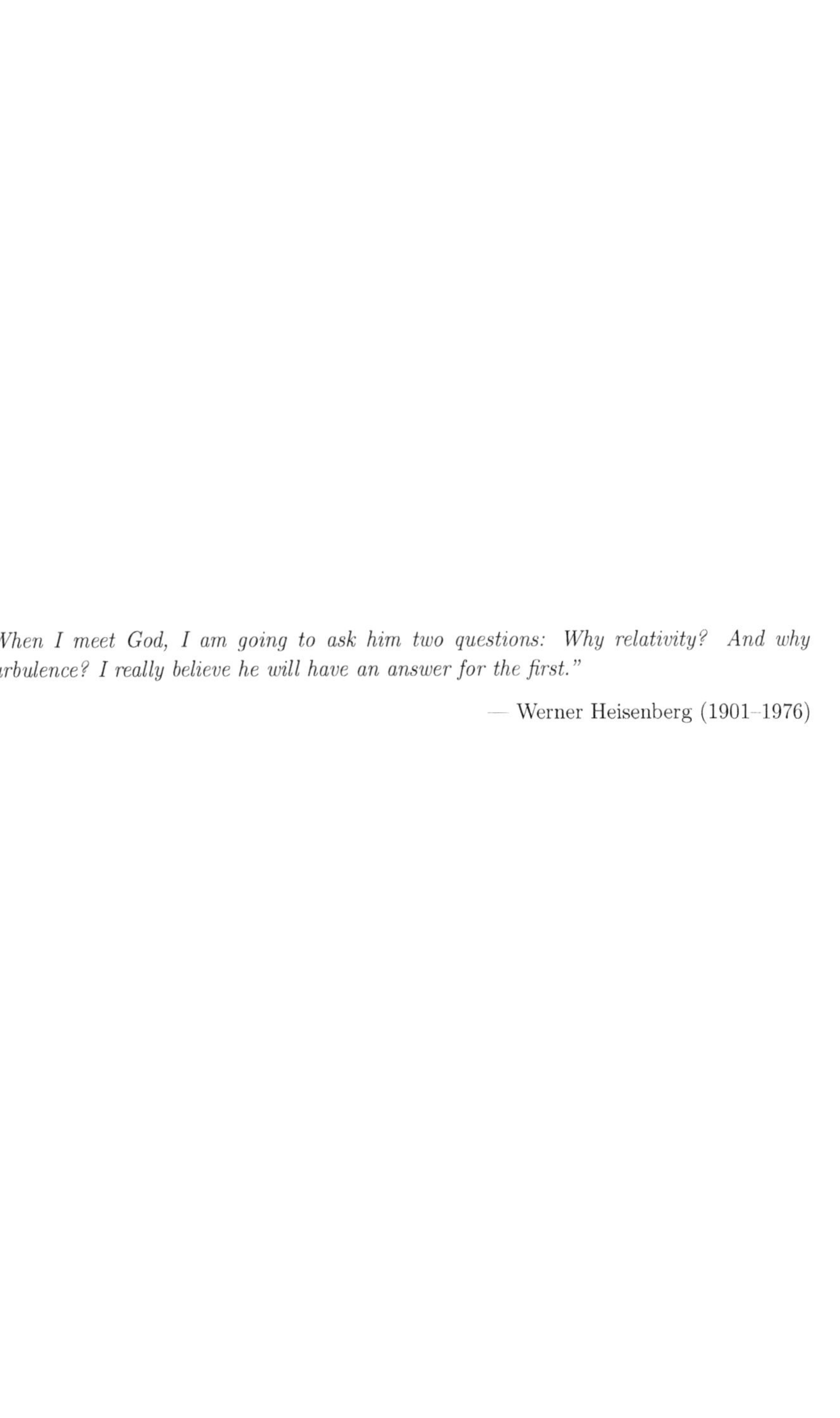

"When I meet God, I am going to ask him two questions: Why relativity? And why turbulence? I really believe he will have an answer for the first."

— Werner Heisenberg (1901–1976)

Abstract

Structured surfaces such as rib turbulators, protrusions, pin fins, or dimples are heat transfer augmentation devices extensively used in heat exchangers. Dimples are already known to promote heat transfer with a minimum hydraulic pressure loss in channel flows. Moreover, different studies published in the literature suspect that dimples induce distinctive unsteady vortex structures able to prevent or mitigate particulate fouling, i.e., the unwanted deposition and accumulation of small suspended particles (e.g., silt or iron oxide) and gravitational settling of larger particles. Fouling, which has been described as one of the major unresolved and most challenging problem in heat transfer, causes an additional thermal resistance as well as an increase of pressure loss and therefore drastically reduces the thermo-hydraulic efficiency of heat exchangers. Main subject of this work is the numerical investigation of dimpled heat transfer surfaces regarding the interaction between unsteady local flow structures, heat transfer and particulate fouling in order to confirm the supposed high fouling-mitigation potential or self-cleaning process.

For the numerical investigations, a new multiphase Eulerian-Lagrangian method, combining scale-resolving large-eddy simulations (LES) with an efficient Lagrangian point-particle tracking (LPT) algorithm, is introduced and implemented into the CFD framework "OpenFOAM". The formation of fouling deposits is primarily based on the DLVO theory and a conversion algorithm, which converts deposited particles into a third phase (i.e., the fouling layer) and deletes them from the costly LPT, additionally enhancing the efficiency of the simulations. The still remaining high computational effort of Eulerian-Lagrangian LES impede the application of the proposed approach for real-time fouling intervals (e.g., hours, days or weeks). To overcome this issue, a multiscale procedure for the simulation of long-term fouling intervals is derived, based on the heterogeneous multiscale method (HMM).

The implemented numerical methods were comprehensively validated using measurements and DNS data for different canonical flows available in the literature, such as a particle-laden turbulent flow through a simplified combustion chamber or a particle-laden turbulent channel flow including particle deposition. Especially the latter one allows the assessment of the Eulerian-Lagrangian approach regarding its ability to capture the important near-wall driving mechanisms in wall-bounded flows, responsible for particle concentration in the near-wall accumulation region.

Subsequent to the validation, an analysis of the thermo-hydraulic efficiency for different clean and fouled structured heat transfer surfaces, namely the square cavity and two spherical dimples with a corresponding dimple depth-to-dimple diameter ratio of $t/D =$

0.26 and 0.35, were performed for different particle mass loadings. These numerical investigations confirm the superior thermo-hydraulic efficiency as well as the high fouling-mitigation potential of dimpled surfaces. The comparison between the measured and simulated mass-based fouling resistance for both dimple configurations highlights the accuracy of the proposed numerical approach. The prediction of particulate fouling on spherical dimples in a staggered arrangement (or dimple package) extends the presented study to an application with higher industrial relevance, since such arrangements are commonly used in many applications such as heat exchangers or gas turbines engines.

Preface

This dissertation is based on the following peer-reviewed papers which have been published during the DFG-funded project "Wirkmechanismen des Partikelfoulings auf strukturierten wärmeübertragenden Oberflächen (effect mechanisms of particulate fouling on structured heat transfer surfaces)".

Paper I

R. Kasper, J. Turnow, and N. Kornev. Numerical modeling and simulation of particulate fouling of structured heat transfer surfaces using a multiphase Euler-Lagrange approach. Int. J. Heat Mass Transfer, 115:932–945, 2017. doi: 10.1016/j.ijheatmasstransfer.2017.07.108.

Paper II

J. Turnow, **R. Kasper**, and N. Kornev. Flow structures and heat transfer over a single dimple using hybrid URANS-LES methods. Comput. Fluids, 172:720–727, 2018. doi: 10.1016/j.compfluid.2018.01.014

Paper III

R. Kasper, H. Deponte, A. Michel, J. Turnow, W. Augustin, S. Scholl, and N. Kornev. Numerical investigation of the interaction between local flow structures and particulate fouling on structured heat transfer surfaces. Int. J. Heat Fluid Flow, 71:68–79, 2018. doi: 10.1016/j.ijheatfluidflow.2018.03.002

Paper IV

R. Kasper, J. Turnow, and N. Kornev. Multiphase Eulerian-Lagrangian LES of particulate fouling on structured heat transfer surfaces. Int. J. Heat Fluid Flow, 79:108462, 2019. doi: 10.1016/j.ijheatfluidflow.2019.108462

Paper V

H. Deponte, **R. Kasper**, S. Schulte, W. Augustin, J. Turnow, N. Kornev, S. Scholl. Local and time resolved investigation of particle deposition on dimpled heat transfer surfaces. Chem. Eng. Sci., 227:115840, 2020. doi: 10.1016/j.ces.2020.115840

Contents

List of Figures

List of Tables

Nomenclature

Roman symbol	Description	Unit
a	Thermal diffusivity	$\mathrm{m^2\,s^{-1}}$
A	Area	$\mathrm{m^2}$
b	Model parameter	–
B	Width	m
$\mathbf{B}$	Body force vector	$\mathrm{m\,s^{-2}}$
c_p	Specific heat capacity	$\mathrm{J\,kg^{-1}\,K^{-1}}$
C	Mass concentration	$\mathrm{kg\,m^{-3}}$
C^+, C_θ^+	Integration constant	–
C_k, C_e	Model constant	–
C_D	Drag coefficient	–
C_f	Fanning friction factor	–
C_K	Kolmogorov constant	–
C_S	Smagorinsky constant	–
$\mathbf{C}$	Cross stress tensor	$\mathrm{m^2\,s^{-2}}$
D	Diameter	m
e	Coefficient of restitution	–
E	Energy	J
	Energy contribution to TKE	$\mathrm{m^2 s^{-2}}$
f	Darcy friction factor	–
	Frequency	Hz
F	Flatness	–
F_1, F_2	Model parameter	–
$\mathbf{F}$	Force	N
G	Filter kernel	–
	Geometry factor	–
h	Enthalpy	J
	Heat transfer coefficient	$\mathrm{W\,m^{-2}\,K^{-1}}$
$\hbar\varpi$	Lifshitz–van der Waals constant	N m
H	Height	m
$\mathbf{I}$	Identity tensor	–
$\mathbf{J}$	Impulsive force	N s
k	Stiffness coefficient	$\mathrm{N\,m^{-1}}$
	Thermal conductivity	$\mathrm{W\,m^{-2}\,K^{-1}}$

	Turbulence kinetic energy	$m^2\,s^{-2}$
K	Permeability	m^2
	Proportionality model constant	–
L	Length scale	m
$\mathbf{L}$	Leonard stress tensor	$m^2\,s^{-2}$
m	Mass	kg
$\dot{m}$	Mass flow rate	$kg\,s^{-1}$
		$kg\,m^{-2}\,s^{-1}$
$\mathbf{M}$	Turbulence stress tensor	m^2s^{-2}
$\mathbf{n}$	Normal vector	–
N	Number	–
p	Yield stress (strength)	$N\,m^{-2}$
	Static pressure	$N\,m^{-2}$
	Specific pressure	$m^2\,s^{-2}$
P, N	Cell-center points	–
$\mathcal{P}$	Turbulence production rate	$m^2\,s^{-3}$
$\mathbf{q}$	Heat flux	$W\,m^{-2}$
Q	Heat flow rate	W
$\mathbf{r}$	Spatial separation vector	m
R	Radius	m
R_f	Fouling resistance	$m^2\,K\,W^{-1}$
$\mathbf{R}$	Cross-correlation function	$m^2\,s^{-2}$
	Reynolds stress tensor	$m^2\,s^{-2}$
S	Distance between dimples	m
	Skewness	–
	Stagnation point	–
	Surface area	m^2
S_ϕ	Source term	–
$\mathbf{S}$	Strain rate tensor	s^{-1}
t	Depth	m
	Time	s
$\mathbf{t}$	Tangential vector	–
T, θ	Temperature	K
T_{lm}	Logarithmic mean temperature	K
$\mathbf{T}$	Stress tensor	$N\,m^{-2}$
u, v, w	Velocity components	$m\,s^{-1}$
$\mathbf{u}$	Velocity vector	$m\,s^{-1}$
U	Internal energy	J
	Overall heat transfer coefficient	$W\,m^{-2}\,K^{-1}$
V	Volume	m^3
V_d	Deposition rate	$m\,s^{-1}$
x, y, z	Cartesian coordinates	m
	Thickness	m
$\mathbf{x}$	Space vector	m

Greek symbol	Description	Unit
α	Phase fraction	–
Γ_ϕ	Diffusion coefficient	$m^2\,s^{-1}$
δ	Overlap distance	m
δt_f	Micro fouling time interval	s
Δ	Difference	e.g., K
Δt_f	Macro fouling time interval	s
$\overline{\Delta}$	Filter width	m
ϵ	Turbulence dissipation rate	$m^2\,s^{-3}$
	Void fraction	–
$\boldsymbol{\zeta}$	Random vector	–
η	Damping coefficient	$kg\,s^{-1}$
	Kolmogorov length scale	m
	Mass loading	–
κ	Von Kármán constant	–
$\boldsymbol{\kappa}$	Wave number tensor	m^{-1}
λ	Free mean path	m
	Interpolation factor	–
μ	Dynamic viscosity	$kg\,m^{-1}\,s^{-1}$
	Friction coefficient	–
ν	Kinematic viscosity	$m^2\,s^{-1}$
ρ	Density	$kg\,m^{-3}$
$\boldsymbol{\rho}$	Auto-correlation function	–
σ	Bulk density	$kg\,m^{-3}$
	Standard deviation	e.g., $m\,s^{-1}$
τ	Characteristic time scale	s
	Shear stress	$N\,m^{-2}$
τ_η	Kolmogorov time scale	s
$\boldsymbol{\tau}_{SGS}$	Subgrid-scale stress tensor	$m^2\,s^{-2}$
ϕ	Flow variable	e.g., K
$\boldsymbol{\Psi}$	Stream function	$m^2\,s^{-1}$
ω	Specific turbulence dissipation rate	s^{-1}
$\boldsymbol{\omega}$	Vorticity vector	s^{-1}
Ω	Physical space	m^3

Dimensionless number	Description
Co	Courant number
Kn	Knudsen number
Nu	Nusselt number

Pr	Prandtl number
Re	Reynolds number
St	Stokes number

Subscript	Description
$(...)_0$	Reference value
$(...)_b$	Bulk
	Value on boundary face
$(...)_c$	Cell value
$(...)_{crit}$	Critical
$(...)_d$	Deposit
$(...)_{def}$	Deformation
$(...)_{dyn}$	Dynamic
$(...)_D$	Based on dimple print diameter
	Downstream
$(...)_{D_h}$	Based on hydraulic diameter
$(...)_{el}$	Elastic
$(...)_E$	Ensemble average
	Explicit
$(...)_f$	Face
	Fluid
	Fouling
$(...)_H$	Based on channel height
$(...)_I$	Implicit
$(...)_{kin}$	Kinetic
$(...)_l$	Loss
$(...)_m$	Mass-based
	Mixture
$(...)_{max}$	Maximal
$(...)_n$	Normal
$(...)_p$	Particle
$(...)_{pl}$	Plastic
$(...)_r$	Removal
$(...)_{rel}$	Relative
$(...)_{rms}$	Root mean square
$(...)_s$	Sample
$(...)_{SGS}$	Subgrid-scale
$(...)_t$	Tangential
	Turbulent
$(...)_{th}$	Thermal
$(...)_{tot}$	Total, overall
$(...)_T$	Temporal average

$(...)_{\tau}$	Based on wall shear stress
$(...)_{U}$	Upstream
$(...)_{vdW}$	Van der Waals
$(...)_{w}$	Wall

Superscript	Description
$(...)^{a}$	Anisotropic part
$(...)^{n}$	Normal direction
$(...)^{t}$	Tangential direction
$(...)^{\mathsf{T}}$	Transpose
$(...)^{*}$	Asymptotic value
$(...)^{+}$	Dimensionless value
$(...)'$	Fluctuation around the mean value
$\lvert ... \rvert$	Magnitude value
$\langle ... \rangle$	Time-averaged mean value
$\overline{...}$	Filtered value
$\widetilde{...}$	Test-filtered value

Abbreviation	Description
BBO	Basset, Boussinesq, Oseen
BC	Boundary condition
BD	Backward differencing
BFS	Backward-facing step
CBB	Confined bluff-body
CD	Central differencing
CFD	Computational fluid dynamics
CFL	Courant-Friedrichs-Lewy
CHT	Conjugate heat transfer
CPU	Central processing unit
CV	Control volume
DEM	Discrete element method
DES	Detached eddy simulation
DLVO	Derjaguin, Landau, Verwey, Overbeek
DNS	Direct numerical simulation
DOE	Dynamic one equation eddy-viscosity model
DPM	Discrete parcel method
DSM	Dynamic Smagorinsky model
EFD	Experimental fluid dynamics
FDM	Finite difference method
FEM	Finite element method
FVM	Finite volume method

GDH	Gradient diffusion hypothesis
HMM	Heterogeneous multiscale method
ICTV	Institute for Chemical and Thermal Process Engineering, Technische Universität Braunschweig
IDDES	Improved delayed detached eddy simulation
LDA	Laser Doppler anemometry
LES	Large-eddy simulation
LPT	Lagrangian particle tracking
MP-PIC	Multiphase Particle-In-Cell method
NVD	Normalized variable diagram
PDE	Partial differential equation
PFQ	Phosphorescent fouling quantification
PIMPLE	Merged PISO-SIMPLE algorithm
PISO	Pressure-implicit with splitting of operators
PSI-Cell	Particle-Source-In-Cell method
QUICK	Quadratic upstream interpolation for convective kinematics
RANS	Reynolds-averaged Navier-Stokes equations
RHH	Reeks, Reed, Hall
SIMPLE	Semi-implicit method for pressure linked equations
SGS	Subgrid scale
SM	Smagorinsky model
SST	Shear stress transport
TFM	Two-fluid model
TKE	Turbulence kinetic energy
TVD	Total variation dimishing
UD	Upwind differencing
(U)RANS	Unsteady Reynolds-averaged Navier-Stokes equations

1 Introduction

1.1 Motivation

The prediction and prevention of fouling is a crucial problem in many industrial processes, such as chemical and process industry, including oil refineries, power generation, energy recovering etc. Fouling, generally defined as the unwanted accumulation of various materials on solid surfaces of processing equipment, causes tremendous problems such as production loss, fuel and maintenance costs [139, 108]. Besides crystallization fouling, particulate fouling due to the deposition of small suspended particles (e.g., clay or iron oxide) and gravitational settling of larger particles is the most important and frequent fouling phenomena [17]. This is especially typical for heat exchangers, shown in Fig. 1.1, illustrating structured surfaces as rib turbulators, pin-fins arrays, protrusions or dimples extensively used to enhance the convective heat transfer [84]. The intensification of heat transfer is achieved by generating secondary flows which interfere the boundary layer growth as well as cause flow recirculation and shear-layer reattachment, promoting mixing and an increase of the turbulence intensity. On the other hand, besides the additional hydraulic losses due to the surface structuring, one of the major disadvantages of structured heat transfer surfaces is the susceptibility for particulate fouling, resulting in a significant reduction of the thermo-hydraulic performance over time.

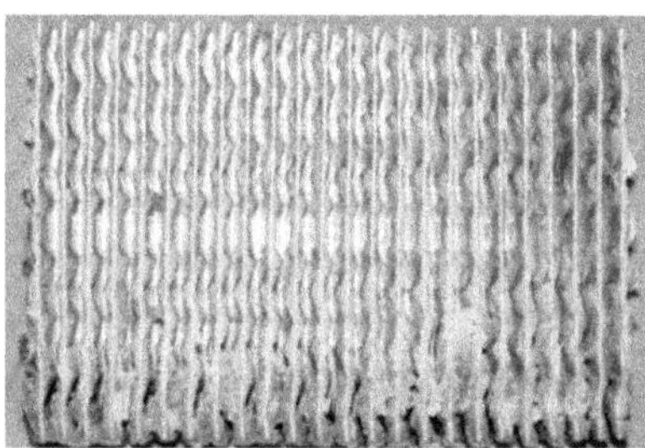

Fig. 1.1: Formation of fouling deposits inside a shell and tube heat exchanger (left) [109]; particulate fouling in a compact heat exchanger for automotive applications (right). Printed by permission from the ICTV, TU Braunschweig.

Various types of structured heat transfer surfaces have been thoroughly investigated with the objective to promote the heat transfer with a minimum hydraulic pressure loss.

Since the application of dimpled surfaces is a very efficient technique to enhance the thermo-hydraulic performance, as shown in Fig. 1.2, they are still primary subject of many experimental and numerical investigations in the field of thermo-fluid dynamics.

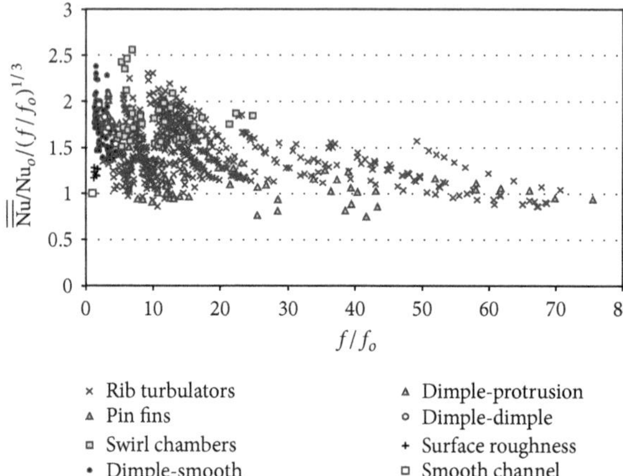

Fig. 1.2: Comparison of the globally averaged thermo-hydraulic performance dependent upon friction factor ratios for various heat transfer enhancement techniques [83].

Experimental studies were performed by Afanasyev et al. [2], who observed a maximum heat transfer augmentation of about 40% accompanied by a low increase of hydraulic losses for a plate with dimples in the turbulent regime, and by Chyu et al. [23], who evaluated the hydraulic losses and heat transfer enhancement for surfaces with an array of hemisphere and tear-drop shaped cavities in the range of Reynolds number, based on the hydraulic diameter of the channel, of $10{,}000 \leq Re_{D_h} \leq 50{,}000$. The results showed that both kinds of concavity configurations induce a heat transfer enhancement up to 2.5 in contrast to the opposite smooth wall whereas the flow resistance was half of that for rib tabulators. The effect of the channel height and dimple depth on heat transfer within the turbulent flow regime has been studied by Mahmood [95] and Ligrani et al. [85, 84]. Comprehensive numerical investigations regarding dimpled surfaces were published by Isaev et al. [62], who performed a study on the influence of the Reynolds number and dimple depth on the turbulent heat transfer and hydraulic loss in a narrow channel using URANS (unsteady Reynolds-averaged Navier-Stokes equations), by Elyyan et al. [42] as well as by Turnow et al. [148, 149]. Especially the numerical study of Turnow [146] and the experimental investigation of Kozlov and Chudnovsky [75] are very interesting, since they mention not merely the superior thermo-hydraulic performance of dimpled surfaces compared to other heat transfer enhancement techniques, but also a possible fouling-mitigation potential or self-cleaning process due to dimples. Kozlov and Chudnovsky explain the fouling mitigation with a tornado-like flow inside the dimple, as illustrated in Fig. 1.3, which directly transports or evacuates approaching fouling particulates out of the dimple back into the core flow. However, it could be shown later by Turnow et al. [148, 146] that tornado-like spatial flow structures can be revealed by the spatial eigenmodes of the velocity field obtained from proper orthogonal decomposition (POD), but they do not exist as pure coherent flow structures. Further current experimental [155, 72, 9] and numerical results [88, 147, 87] show present significance of fundamental studies exploring advantages and disadvantages of dimpled surfaces.

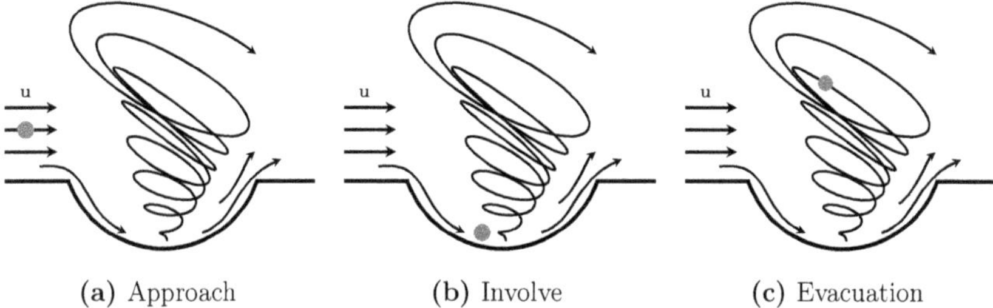

(a) Approach (b) Involve (c) Evacuation

Fig. 1.3: Mechanism of fouling mitigation through dimpled surfaces, according to Kozlov and Chudnovsky [75].

In contrast to this, investigations of structured heat transfer surfaces considering particulate fouling are relatively seldom. Available studies of particle-laden flows over non-flat surfaces, which emphasize the influence of surfaces curvature on the particle statistics as well as on particle accumulation and deposition, are presented by Marchioli et al. [97], Milici et al. [105] or De Marchis et al. [29], who investigated the influence of corrugated surfaces or roughnesses on particle dynamics in turbulent channel flows. A recent study is presented by Luo et al. [92], who analyzed effect of staggered arranged hemispherical roughness elements on a turbulent particle-laden flow. The most fundamental fouling modeling approach was proposed by Kern and Seaton [69], modeling the fouling processes as a balance between deposition and removal process. Further fouling models are described by Suitor et al. [141], Epstein [43], Bohnet [15] and Bott [17], but they allow only an integral fouling evaluation without accounting for local flow features and critical flow conditions (e.g., hot spots or low velocity zones). Latest numerical contributions are from Tong et al. [144], who simulated two-dimensional particle deposition and removal processes on tubes by coupling a multiple-relaxation-time lattice Boltzmann method with a finite volume procedure, and Wang et al. [159, 160] with a parameter study on the fouling characteristic of H-type finned heat exchanger using a two-dimensional RANS approach.

1.2 Fouling of heat transfer surfaces

The reasons and mechanisms of deposition formations on heat transferring surfaces are extremely manifold. Thus, the different types of fouling are usually subdivided into five main categories based on the underlying key physical/chemical processes [44, 15, 17, 108]:

Crystallization fouling Precipitation and deposition of dissolved salts, which at process conditions become supersaturated at the heat transfer surface or solidification fouling due to the shortfall of the solidification temperature of a dissolved component (e.g., solidification of wax from crude oil streams).

Particulate fouling Deposition of small suspended particles on heat transfer surfaces of any orientation and/or gravitational settling of larger particles onto horizontal surfaces.

Chemical reaction fouling Deposition formations at heat transfer surfaces by a chemical reaction, whereby the surface material itself is not part of the chemical reaction.

Corrosion fouling Formation of a corrosion layer on heat transfer surfaces which usually causes a low (additional) thermal resistance due the relatively high thermal conductivity of oxides. However, the enhanced surface roughness may promote other kinds of fouling.

Biological fouling Development and deposition of organic films consisting of microorganisms and their products such as bacteria and the attachment and growth of macroorganisms (e.g., mussels, algae) on heat transfer surfaces.

In practice or rather under real conditions several fouling mechanisms appear simultaneously, nearly always being mutually reinforcing. One exception is the combination of particulate and crystallization fouling, where particles of the crystallizing matter accelerate fouling, whereas particles from the other material may lead to reduced fouling due to a weakening of the deposit formation [108]. All mentioned fouling mechanisms occur in the following five consecutive steps [44, 15, 108]:

Initiation period The initially high overall heat transfer coefficient of a new or cleaned heat exchanger often remains unchanged for a certain time (i.e., no fouling occurs). The duration of this initiation period depends on various parameters, e.g., temperature and surface roughness.

Mass transport Transport of at least one (fouling) key component to the heat transfer surface through the fluid bulk, which is mostly achieved by diffusion. In case of particle transport to the surface, the consideration of inertia effects as well as thermophoretic and turbophoretic (only for turbulent carrier flows) forces is required.

Attachment and formation As soon as the transport to the heat transfer surface is completed, the foulant must stick to the heat transfer surface (for particulate fouling) or react to the deposit forming substance (e.g., $CaCO_3$).

Removal Depending on the strength of the deposit, erosion or removal due to the fluid flow (i.e., shear stresses acting on the surface of the fouling layer) occurs immediately after the first formation of deposits.

Aging The strength of the fouling layer can change in time. This aging process is referred to as *aging* and can either increase or decrease the strength of the deposits.

1.2.1 Fouling resistance and fouling curves

Fouling on heat transfer surfaces is generally considered in the design process of heat exchangers by using the so-called *thermal fouling resistance* $R_{f,th}$ in the evaluation of the overall heat transfer coefficient U [109]. For a clean heat transfer surface, the heat flow resistance is defined as:

$$\frac{1}{U_0} = \frac{1}{h_1} + \frac{x_w}{k_w} + \frac{1}{h_2} = \frac{1}{h_1} + R_w + \frac{1}{h_2}, \tag{1.1}$$

in which h_1 and h_2 are the corresponding heat transfer coefficients of both heat exchanging fluids, x_w and k_w is the thickness and the thermal conductivity of the heat transfer surface (or wall), respectively, and R_w is the thermal resistance of the separating wall. Assuming that fouling occurs only on one side of the heat transfer wall, which is valid for many industrial applications, the overall heat flow resistance of the fouled surfaces is [110]:

$$\frac{1}{U_f} = \frac{1}{h_1} + \frac{x_w}{k_w} + \frac{1}{h_2} + \frac{x_f}{k_f} = \frac{1}{h_1} + R_w + \frac{1}{h_2} + R_{f,th}, \tag{1.2}$$

where x_f is the thickness of the fouling layer, k_f is the thermal conductivity of the foulant material and $R_{f,th}$ is the fouling resistance, which can be interpreted as additional thermal resistance due to fouling. For constant flow rates, the fouling resistance at any time is the difference between the actual heat transfer resistance and the initial (clean) heat transfer resistance:

$$R_{f,th} = \frac{1}{U_f(t)} - \frac{1}{U_0}. \tag{1.3}$$

The time-dependent overall heat transfer coefficient is calculated from

$$U(t) = \frac{Q(t)}{A \cdot \Delta T_{lm}(t)}, \tag{1.4}$$

with the logarithmic mean temperature difference ΔT_{lm} (see e.g., Incropera et al. [61]) and the heat flow rate Q, which is given by:

$$Q = \dot{m} \cdot c_p \cdot \Delta T, \tag{1.5}$$

where ΔT is the temperature difference of either the heating or cooling fluid, $\dot{m}$ is the mass flow rate and c_p is the corresponding heat transfer capacity. The thermal fouling resistance can be easily monitored without any interruption of the process but allows only an assessment of the integral fouling behavior, since the temperatures are only measured at the inlet and outlet for the heat exchanging fluids. However, investigations of operating industrial heat exchanger have shown that fouling often follows a decreasing or even asymptotic trend. Based on this observation, Kern and Seaton suggested modeling the fouling process as a balance between competing transport processes to and from the heat transfer surface, namely, deposition and removal [109]. Therefore, the accumulation of the deposited fouling mass m_f per unit area[1] in time t is expressed as [69]:

$$\frac{\mathrm{d}m_f}{\mathrm{d}t} = \dot{m}_d - \dot{m}_r = \frac{\mathrm{d}R_{f,th}}{\mathrm{d}t}\rho_f k_f, \tag{1.6}$$

assuming that the thermal conductivity k_f and the density ρ_f of the fouling material remain constant with time and deposit thickness. Based on the Kern and Seaton model, Eq. (1.6), the *fouling curves* for three essentially different cases are shown in Fig. (1.4).

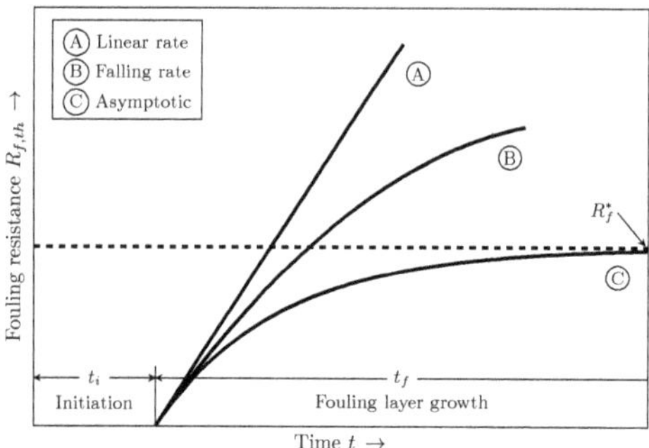

Fig. 1.4: Possible dependence of the fouling resistance $R_{f,th}$ on time t: idealized thermal fouling curves for a linear, falling and asymptotic growth rate.

Curve A shows a linear increase of the fouling resistance with time. This behavior is mainly observed when no removal of solids (i.e., $\dot{m}_r = 0\,\mathrm{kg/m^2s}$) takes place. For curve B, a slowing down (or falling rate) of the solids increase with increasing fouling layer thickness is observed, but without reaching a maximum. Curve C shows the characteristic asymptotic fouling behavior, which is , after a certain time, the solid removed per unit time and surface area is equal to the deposited solid (i.e., $\dot{m}_d - \dot{m}_r = 0\,\mathrm{kg/m^2s}$). Moreover, the mass of the fouling layer per unit area and thus also the fouling resistance reach limiting values, the so-called *asymptotic fouling resistance* $R^*_{f,th}$. Thereby, it is possible that, during the initial phase of fouling, the fouling resistance shows a linear increase [15].

1.2.2 Modeling of particulate fouling

Existing physically based models for the prediction of particulate fouling (see e.g., Müller-Steinhagen [109] or Bohnet [15]) are usually derived under the assumption that the growth of the fouling layer is leveled after a certain time. Following the pioneering work of Kern and Seaton [69], the progression of the fouling resistance with time can be expressed as:

$$\frac{\mathrm{d}R_{f,th}}{\mathrm{d}t}\rho_f k_f = \dot{m}_d - \dot{m}_r. \tag{1.7}$$

[1]Please note that the fouling mass m_f, the deposition rate $\dot{m}_d$ as well as the removal rate $\dot{m}_r$ are expressed as mass or mass flow rate per unit area (i.e., in $\mathrm{kg/m^2}$ and $\mathrm{kg/m^2s}$).

According to Müller-Steinhagen [109], the deposition rate $\dot{m}_d$ can be modeled through a simplified mass transfer correlation as being proportional to the bulk flow velocity u and the foulant concentration c:

$$\dot{m}_d = K_1 u c, \tag{1.8}$$

where K_1 is a (proportionality) model constant. Since the removal of deposited fouling material is mainly caused by shear forces from the bulk flow, the removal rate $\dot{m}_r$ is derived under the assumption that it is proportional to the shear stress τ_f acting on the surface of the fouling layer and to the thickness x_f of the fouling layer:

$$\dot{m}_r = K_2 \tau_f x_f, \tag{1.9}$$

with model constant K_2. Both modeling approaches are based on simplistic assumptions and ignore several mechanism that may be responsible for accumulation of fouling deposits on heat transfer surfaces and have therefore been frequently criticized, extended and improved. However, combining Eqs. (1.8) and (1.9) and integration with respect to time yields:

$$R_{f,th}(t) = \frac{K_1 u c}{\rho_f k_f K_2 \tau_f}\left(1 - e^{-K_2 \tau_f t}\right) = R^*_{f,th}\left(1 - e^{-bt}\right), \tag{1.10}$$

which includes the asymptotic fouling resistance $R^*_{f,th}$. This fundamental relationship describes the increase of the fouling resistance in time, which approaches the limiting value asymptotically. It is obvious that the final value of the fouling resistance increases with the foulant concentration in the carrier flow and decreases with the flow velocity, since the shear stress acting on the surface of the fouling layer is proportional to the square of the bulk flow velocity (i.e., $\tau_f \propto \rho u^2$). Moreover, Eq. (1.10) states that until the asymptotic fouling resistance is attained, the increase of the fouling resistance depends only on the bulk flow velocity as shown in Fig. 1.5.

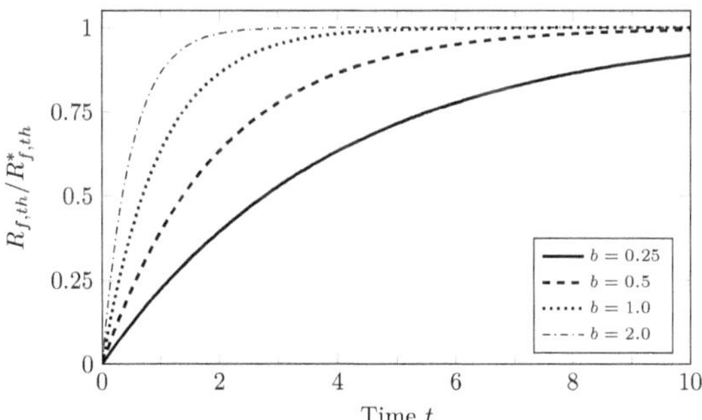

Fig. 1.5: Fouling resistance $R_{f,th}/R^*_{f,th}$ as a function of time t and parameter b in Eq. (1.10), according to the Kern and Seaton model [69].

Although the presented modeling approach, Eq. (1.10), provides essential information regarding the fouling process and reveals fundamental physical dependencies, it suffers from the drawback that it allows only a pure integral evaluation and assessment of particulate fouling without accounting for local flow features, e.g., switching vortices caused by dimples. Furthermore, a precise calibration of the model constants, which is based on experimental fouling investigations, is necessary to achieve reliable results.

1.3 Objectives and thesis outline

A careful review of the available literature has shown, that no numerical investigations of dimpled heat transfer surfaces regarding the interaction between local flow structures, convective heat transfer and fouling deposits using transient large-scale resolving numerical methods (e.g., LES or IDDES) exist at this moment. Moreover, the supposed fouling-mitigation potential of dimpled surfaces is still not confirmed numerically. This work is aimed to fill up this lack of knowledge by introducing a new multiphase Eulerian-Lagrangian approach which is suitable for CFD studies of heat transfer enhancement methods under consideration of particulate fouling using large-scale resolving methods, providing the opportunity to analyze the interaction between local flow structures and different fouling processes in a more comprehensive way. Thus, the two main objectives of this thesis are:

1. Development of a novel numerical approach for spatial- and time-resolved simulations of particulate fouling on structured heat transfer surfaces, based on Eulerian-Lagrangian LES. Extensive validation of the presented multiphase Eulerian-Lagrangian approach for canonical test cases (e.g., particle-laden backward-facing step flow, particle-laden channel flow).

2. Investigation of dimpled heat transfer surfaces regarding their ability to mitigate particulate fouling, while varying the dimple geometry (depth-to-dimple diameter ratio) and dimple arrangement (single spherical dimple and spherical dimples in a staggered arrangement, i.e., dimple package) as well as the carrier flow conditions (bulk flow velocity and mass loading of the dispersed phase).

This dissertation is divided into two parts. The first part contains the theoretical background including the derivation of the governing equations for the continuous phase (carrier flow) and the dispersed phase (foulant material), the numerical methodology as well as a description of the implemented fouling model. The second part presents the validation of the numerical methods and the results of the performed numerical fouling investigations for dimpled heat transfer surfaces followed by a conclusion.

2 Description of the continuous phase

In this chapter, the fundamental equations of fluid motions and heat transfer are derived, allowing a full description of the investigated incompressible carrier flows (i.e., continuous phase). Additionally, a condensed introduction into turbulent wall-bounded flows and different mathematical turbulence modeling approaches, with emphasis on the large-eddy simulation (LES), is given. More information and detailed discussions on fluid dynamics, turbulence, and modeling approaches for turbulent flows can be found in [89, 116, 123, 82, 124, 10].

2.1 Governing equations of fluid motion

The governing equations of fluid dynamics are derived under continuum assumption, which means that the considered fluid is a continuum state of matter and all flow variables can be defined at any point. This assumption holds for very small Knudsen numbers

$$Kn = \frac{\lambda}{L} \ll 1, \tag{2.1}$$

or more precisely, if the characteristic length scale of the flow L is much larger than the mean free path of the flow particles λ. In this case, the fluid motion is described through the principle of conservation of mass, momentum and energy expressed in a continuous way in space $\mathbf{x}$ and time t.

2.1.1 Continuity equation

Mass conservation implies that the mass of a fluid flow remains constant. Hence, for a fixed fluid-containing control volume $\mathrm{d}V$, the difference of the mass flow, which enters and leaves the considered control volume through its boundaries, is zero. This relation can be formulated using the following integral equation:

$$\frac{\partial}{\partial t}\int_V \rho \,\mathrm{d}V + \int_S \rho \mathbf{u} \cdot \mathbf{n}\,\mathrm{d}S = 0, \tag{2.2}$$

where ρ is the fluid density, $\mathbf{u}$ is the fluid velocity, $\mathrm{d}S$ is the infinitesimally small surface area and $\mathbf{n}$ is the unit vector orthogonal to $\mathrm{d}S$. The integral mass conservation, Eq. (2.2), can be transformed into the differential vector form by applying Gauss' divergence theorem, see Eq. (4.6), to the convection term and introducing the del operator ∇:

$$\frac{\partial \rho}{\partial t} + \nabla \cdot (\rho \mathbf{u}) = 0. \tag{2.3}$$

For incompressible or density-constant flows (i.e., flows in which ρ is independent both of space and time), as considered in this work, this equation reduces to:

$$\nabla \cdot \mathbf{u} = 0. \tag{2.4}$$

Thus, the mass conservation is satisfied for incompressible flows with solenoidal (source- and sink-free) or divergence-free velocity fields.

2.1.2 Momentum equation

The momentum balance equation is based on Newtons second law of motion, stating that the temporal change of momentum inside a fixed fluid-containing volume of space $\mathrm{d}V$ is equal to the sum of all surface forces (e.g., pressure, normal and shear stresses) and body forces (e.g., gravity, centrifugal and Coriolis force) acting on the fluid in the considered control volume [46]. Thus, the integral form of the momentum equation becomes:

$$\frac{\partial}{\partial t} \int_V \rho \mathbf{u} \, \mathrm{d}V + \int_S \rho \mathbf{u}\mathbf{u} \cdot \mathbf{n} \, \mathrm{d}S = \int_S \mathbf{T} \cdot \mathbf{n} \, \mathrm{d}S + \int_V \rho \mathbf{B} \, \mathrm{d}V, \tag{2.5}$$

with the stress tensor $\mathbf{T}$, which is the molecular rate of transport of momentum, and vector $\mathbf{B}$, including the body forces (per unit mass). The surface integrals (convective and diffusive fluxes) need to be converted into volume integrals by applying Gauss' divergence theorem, which leads to following differential vector form of the momentum balance equation (Euler's first law of motion):

$$\frac{\partial \rho \mathbf{u}}{\partial t} + \nabla \cdot (\rho \mathbf{u}\mathbf{u}) = \nabla \cdot \mathbf{T} + \rho \mathbf{B}. \tag{2.6}$$

For Newtonian fluids the general stress tensor $\mathbf{T}$ is expressed by the static pressure and shear stresses, which results in:

$$\mathbf{T} = -\left(p + \frac{2}{3}\mu \nabla \cdot \mathbf{u}\right)\mathbf{I} + 2\mu \mathbf{S}, \tag{2.7}$$

where μ is the dynamic viscosity, $\mathbf{I}$ is the identity tensor, p is the static pressure and $\mathbf{S}$ is the strain rate tensor:

$$\mathbf{S} = \frac{1}{2}\left(\nabla \mathbf{u} + \nabla \mathbf{u}^{\mathsf{T}}\right). \tag{2.8}$$

For incompressible Newtonian fluids, the stress tensor can be further simplified due to mass conservation (Eq. (2.4)):

$$\mathbf{T} = -p\mathbf{I} + 2\mu\mathbf{S}. \tag{2.9}$$

Substitution of this simplified stress tensor for incompressible Newtonian fluids into the vector form of the momentum balance equation, Eq. (2.6), and division by the fluid density ρ leads to:

$$\frac{\partial \mathbf{u}}{\partial t} + \boldsymbol{\nabla} \cdot (\mathbf{u}\mathbf{u}) = -\frac{1}{\rho}\boldsymbol{\nabla} p + \nu \nabla^2 \mathbf{u} + \mathbf{B}. \tag{2.10}$$

The continuity and momentum equation, Eqs. (2.4) and (2.10), build up a system of partial differential equations (PDE's), the well-known *Navier-Stokes* equations for incompressible flows, named after the French engineer and physicist Claude-Louis Navier (1785-1836) and the Irish mathematician and physicist Sir George Gabriel Stokes (1819-1903).

2.1.3 Energy equation

For the derivation of the energy equation, the convective and diffusive transport as well as the different forms of energy (internal and kinetic energy) have to be considered. Additionally, multiple types of energy sources (e.g., thermal radiation, exothermic reactions) need to be taken into account. Commonly, the contribution of the specific kinetic energy, stress and volume forces is negligible small compared to the internal energy [115]. In this case, the integral form of the energy balance for a fixed fluid-containing volume of space $\mathrm{d}V$ is expressed by the first law of thermodynamics:

$$\frac{\partial}{\partial t}\int_V \rho U \,\mathrm{d}V + \int_S \rho\left(U + \frac{p}{\rho}\right)\mathbf{u} \cdot \mathbf{n}\,\mathrm{d}S = -\int_S \mathbf{q} \cdot \mathbf{n}\,\mathrm{d}S + \int_V S_H \,\mathrm{d}V. \tag{2.11}$$

where U stands for the internal energy, $\mathbf{q}$ for the heat flux over the control volume boundaries due to heat conduction, S_H for the source term which accounts for an additional energy input into the control volume (e.g., absorption of thermal radiation) and $h = U + p/\rho$ for the enthalpy. Assuming an incompressible fluid with small pressure variations (i.e., isochoric and isobaric heat capacities are equal), the simplified caloric equation of state reads:

$$\mathrm{d}U = \mathrm{d}h = c\,\mathrm{d}T. \tag{2.12}$$

Recalling that the heat flux is given by Fourier's law

$$\mathbf{q}(\mathbf{x}) = -k\boldsymbol{\nabla} T(\mathbf{x}) \tag{2.13}$$

and that the thermal diffusivity is $a = k/(\rho c_p)$, the integral energy balance equation, Eq. (2.11), can be rewritten by applying Gauss' divergence theorem and neglecting outer heat sources:

$$\frac{\partial T}{\partial t} + \boldsymbol{\nabla} \cdot (\mathbf{u}T) = \frac{\nu}{Pr}\nabla^2 T, \tag{2.14}$$

whereas Pr is the Prandtl number, which is defined as the ratio of momentum diffusivity (kinematic viscosity) ν to thermal diffusivity a:

$$Pr = \frac{\nu}{a}. \tag{2.15}$$

Eq. (2.14) represents the temperature form of the simplified energy equation (incompressible fluid, small pressure variations) and is derived for a constant thermal diffusivity. This formulation is similar to a general scalar transport equation, whereas the terms on the left-hand side describes the temperature change in time and the heat transfer through convection. The right-hand side accounts for the heat diffusion due to conduction. Since the impact of the temperature T on the thermophysical properties is neglected ($a \approx$ const), the energy equation is fully decoupled from the Navier-Stokes equations and can be solved independently.

2.2 Turbulence in wall-bounded flows

The majority of fluid flows occurring in nature or engineering applications is turbulent. Turbulence is all around us, yet it is usually invisible. Important examples are external flows over all kinds of vehicles such as airplanes and ships, or geophysical flows (rivers, atmospheric boundary layer). Surprisingly, even today physicists still do not agree on how to define turbulence which makes it to one of the oldest unsolved problems in physics. To overcome this issue, turbulence is characterized by specific features such as irregularity, diffusivity, rotationality, dissipation, three-dimensionality and nonlinearity [89, 58, 78].

The most important fundamental relation to evaluate the turbulence intensity within a fluid is the ratio of the inertial forces to viscous forces. Although this concept was originally introduced by George Stokes in 1851 [140], this relation is know as a dimensionless quantity named after Osborne Reynolds (1842–1912), who was the first who studied the conditions for transition to turbulence in pipe flows. The mathematical definition of this so-called *Reynolds number* reads:

$$Re = \frac{u_0 l_0}{\nu}, \tag{2.16}$$

where u_0 and l_0 are the reference velocity and reference length and ν is the kinematic viscosity. For small Reynolds numbers ($Re \sim 1$), the fluid motion is dominated by viscous forces which leads to a laminar flow, whereas for high Reynolds numbers ($Re \gg 1$), the fluid motions is governed by inertial forces resulting in a turbulent flow. Due to the random and unpredictable character of turbulence, an exact prediction of the instantaneous velocity field $\mathbf{u}(\mathbf{x}, t)$ is not possible. However, turbulent flows can be described extensively through statistical quantities (e.g., averaged values), which requires the Reynolds decomposition

[121] of the velocity field into its mean value $\langle \mathbf{u}(\mathbf{x},t)\rangle$ and the time varying fluctuation $\mathbf{u}'(\mathbf{x},t)$:

$$\mathbf{u}'(\mathbf{x},t) = \mathbf{u}(\mathbf{x},t) - \langle \mathbf{u}(\mathbf{x},t)\rangle, \tag{2.17}$$

whereas

$$\langle \mathbf{u}'(\mathbf{x},t)\rangle = 0. \tag{2.18}$$

To extract the statistics of the flow, the statistical average $\langle \mathbf{u}(\mathbf{x},t)\rangle$ has to be evaluated by appropriate averaging operations. The simplest way to achieve this is to apply the *ensemble averaging*, which is defined as the statistical average over N repetitions of an experiment with identical boundary conditions, starting from initial conditions as close as possible:

$$\langle \mathbf{u}(\mathbf{x},t)\rangle_E = \lim_{N\to\infty} \sum_{n=1}^{N} \mathbf{u}^{(n)}(\mathbf{x},t). \tag{2.19}$$

Whereas for steady flows, it is common to introduce the *temporal average* defined as:

$$\langle \mathbf{u}(\mathbf{x},t)\rangle_T = \lim_{T\to\infty} \frac{1}{T} \int_0^T \mathbf{u}(\mathbf{x},t)\,\mathrm{d}t\,. \tag{2.20}$$

Besides the mean value $\langle \mathbf{u}(\mathbf{x},t)\rangle$, high-order moments such as variance, skewness and flatness can be used for a statistical description of turbulent flows (see A.1). In addition, one can introduce correlation functions in case of statically stationary processes. The so-called *cross-correlation* function for the velocity fluctuations at two different points and time instants is given by:

$$\mathbf{R}(\mathbf{r},\tau) = \langle \mathbf{u}'(\mathbf{x},t)\mathbf{u}'(\mathbf{x}+\mathbf{r},t+\tau)\rangle. \tag{2.21}$$

Normalization of Eq. (2.21) for two different points but at the same time ($\tau = 0$) yields the *auto-correlation* function in space:

$$\boldsymbol{\rho}(\mathbf{r},t) = \frac{\langle \mathbf{u}'(\mathbf{x},t)\mathbf{u}'(\mathbf{x}+\mathbf{r},t)\rangle}{\langle \mathbf{u}'(\mathbf{x},t)\mathbf{u}'(\mathbf{x},t)\rangle}. \tag{2.22}$$

Analogously, the auto-correlation function in time ($\mathbf{r} = 0$) is defined as:

$$\boldsymbol{\rho}(\mathbf{x},\tau) = \frac{\langle \mathbf{u}'(\mathbf{x},t)\mathbf{u}'(\mathbf{x},t+\tau)\rangle}{\langle \mathbf{u}'(\mathbf{x},t)\mathbf{u}'(\mathbf{x},t)\rangle}. \tag{2.23}$$

Eq. (2.22) is often used to ensure whether the computational domain is large enough to capture all existing flow structures. Moreover, under the assumption of homogeneous turbulence, it is possible to transform the velocity auto-correlation function into the Fourier space, allowing a more descriptive representation and analysis of the energy processes in turbulent flows by derivation of the spatial energy spectrum $E(\kappa)$ of the velocity fluctuations [116]:

$$E(\kappa) = \iiint\limits_{-\infty}^{+\infty} \frac{\mathbf{\Phi}(\boldsymbol{\kappa})}{2} \delta(\|\boldsymbol{\kappa}\| - \kappa) \, \mathrm{d}\boldsymbol{\kappa} \,, \tag{2.24}$$

where $\mathbf{\Phi}(\boldsymbol{\kappa})$ is the Fourier mode tensor and $\boldsymbol{\kappa}$ is the wave number tensor. The one-dimensional spatial energy spectrum $E(\kappa)$ and its three main ranges, representing the so-called *energy cascade* (i.e., the energy transport from large to small scales of turbulent motion), is illustrated in Fig. (2.1) in the logarithmic scale.

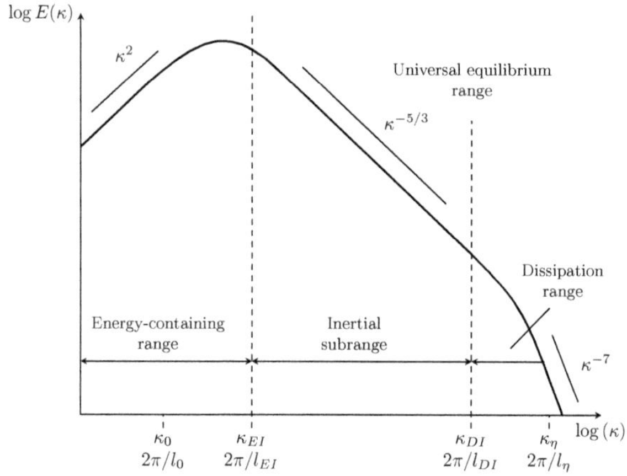

Fig. 2.1: One-dimensional spatial energy spectra $E(\kappa)$ in isotropic turbulence.

The *energy-containing range* $(0 \leq \kappa \leq \kappa_{EI})$ is dominated by anisotropic large-scale fluid motions, carrying the major portion of the turbulent flow due to outer forces. The intermediate range between the energetic eddies and the dissipative small-scale structures is the *inertial subrange* $(\kappa_{EI} \leq \kappa \leq \kappa_{DI})$ in which the energy decay is characterized by $E(\kappa) \sim \kappa^{-5/3}$, also known as Kolmogorov's 5/3 law. The fluid scales in the inertial subrange are independent of outer forces and are dominated by inertial rather than viscous forces. The energy of the flow is transferred from the large scales to the very small ones at which the energy further dissipates due to viscosity. Thus, turbulent motions are damped by viscous effects resulting in a strong decay of kinetic energy at high wave numbers. The *dissipation range* $(\kappa_{DI} \leq \kappa)$ contains the small vortices (down to the *Kolmogorov length scale*[1]) which are independent of outer forces and the flow configuration (e.g., geometry,

[1]The Kolmogorov length scale $\eta = \left(\nu^3/\epsilon\right)^{1/4}$ is the smallest length scale in a turbulent flow and only depend on the dissipation rate ϵ as well as on the kinematic viscosity ν. The corresponding Kolmogorov time scale is defined as $\tau_\eta = (\nu/\epsilon)^{1/2}$.

inflow conditions) and can be assumed as locally isotropic. In 1941, Kolmogorov postulated that at sufficiently high Reynolds numbers *Re* the form of the energy spectrum is universal and the flow only depends on its kinematic viscosity ν and turbulence dissipation rate ϵ [73]. From similarity analysis it was shown, that the energy spectrum $E(\kappa)$ reads:

$$E(\kappa) = C_K \epsilon^{2/3} \kappa^{-5/3}, \tag{2.25}$$

where C_K is the *Kolmogorov constant* which is estimated from experiments and simulations as $C_K = 1.4 - 1.5$. Furthermore, Kolmogorov claimed that the production of turbulence kinetic energy happens on length scales much larger than the dissipative length scales. This assumption is valid for high-Reynolds number flows and was confirmed for a wide range of turbulent flows [132]. However, it should be mentioned that a transfer of energy from small to large fluid scale is possible. This effect occurs due to reunion of small eddies and is known as *backscatter* of turbulence kinetic energy.

Strictly speaking, the aforementioned considerations regarding the energy cascade in turbulent flows are only valid for isotropic and anisotropic turbulence at high Reynolds numbers ($Re \gg 1$), thus without influence of solid walls or at least far away from it, since the Reynolds number close to walls is small ($Re \sim 1$). Moreover, the growth of vortices within the near-wall region is spatially restricted by the wall. Consequently, the corresponding energy spectrum shows no inertial subrange [70]. Following Schlichting's boundary-layer theory [127], two different flow regimes (two-layer structure) occur in *wall-bounded flows* (e.g., pipe and channel flows) due the presence of wall shear stress τ_w (i.e., shear force per unit area) defined as:

$$\tau_w = \rho \nu \left. \frac{\mathrm{d}u}{\mathrm{d}y} \right|_w, \tag{2.26}$$

where u is the stream-wise velocity and y is the wall-normal coordinate. On the one hand a large *core layer* (or outer layer) where the molecular momentum transfer can be neglected compared to the turbulent momentum transfer, and on the other hand a *thin wall layer* (inner layer) where both turbulent and molecular momentum transfer act. In order to simplify the comparison of experimental and numerical results for different geometries and flow configurations, all important flow variables are usually expressed in terms of the (wall) friction velocity

$$u_\tau = \sqrt{\frac{\tau_w}{\rho}}, \tag{2.27}$$

where the wall shear stress τ_w follows directly from Eq. (2.26). Based on the friction velocity u_τ, the dimensionless velocity u^+ and the dimensionless coordinate (or inner coordinate) y^+ in wall-normal direction can be expressed through:

$$u^+ = \frac{u}{u_\tau}, \qquad y^+ = \frac{y u_\tau}{\nu}. \tag{2.28}$$

The universal velocity profile $u^+\left(y^+\right)$, often referred to as *"law of the wall"*, and the universal temperature profile $\theta^+\left(y^+\right)$ for a fully developed turbulent channel flows is shown in Fig. (2.2).

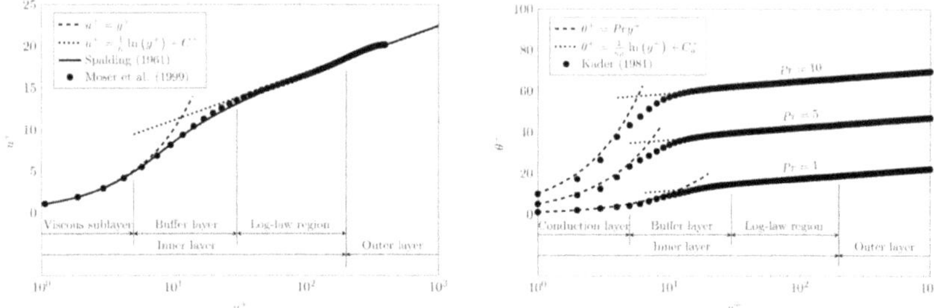

Fig. 2.2: Mean velocity and temperature profiles within a fully developed turbulent channel flow (*"law of the wall"*): universal velocity profile u^+ in inner coordinates (left) compared to Spalding's function (1961) [137] and DNS results published by Moser et al. (1999) [107]; universal temperature profile θ^+ in inner coordinates (right) for various Prandtl numbers compared to Kader's empirical correlation (1981) [65].

The universal character of the mean velocity profile $u^+\left(y^+\right)$ could be derived theoretically based on an analysis of the turbulent Couette flow [127] and is approved by experimental [40] and numerical investigations [71]. It can be distinguished between the outer layer and the inner layer including the viscous sublayer as well as the buffer layer and the log-law region. The *viscous sublayer* ranges from $0 < y^+ < 5$ and is primary dominated by the viscosity of the fluid whereas turbulent fluctuations are negligible. This region can be approximately described by a linear relation between the dimensional velocity u^+ and the dimensionless wall distance y:

$$u^+ = y^+. \tag{2.29}$$

In the *buffer layer* $\left(5 < y^+ < 30\right)$, in which viscous and turbulent stresses are in equilibrium, the dependence of u^+ on y^+ changes from linear to logarithmic variation. The *log-law region* extends from $y^+ > 30$ and is dominated by effects of inertia forces. The velocity profile $u^+\left(y^+\right)$ is prescribed by the logarithmic law of the wall (or simply, the *"log-law"*) [116]:

$$u^+ = \frac{1}{\kappa}\ln\left(y^+\right) + C^+, \tag{2.30}$$

where $\kappa \approx 0.4$ is the *von Kármán constant* and C^+ is an integration constant which is estimated through measurements and simulations for a smooth wall as $C^+ \approx 5 - 5.5$ [116, 79, 146]. The same considerations as for the flow velocity are valid for the temperature, which is also prescribed by a mean profile, Fig. 2.2, depending on heat conduction directly at the wall whereas convection processes are dominating in the outer layer of the flow. Analogously to the dimensionless velocity u^+, a dimensionless temperature can be introduced as:

$$\theta^+ = \frac{\overline{T}_w - T}{T_\tau}, \tag{2.31}$$

in which $\overline{T}_w$ is the time-averaged wall temperature and T_τ is the friction temperature, evaluated based on Eqs. (2.13) and (2.15):

$$T_\tau = \frac{\dot{q}_w}{\rho c_p u_\tau} = -\frac{\nu}{Pr}\frac{1}{u_\tau}\frac{\partial T}{\partial y}\bigg|_w. \tag{2.32}$$

It follows directly from Eq. (2.32) that the mean temperature profile $\theta^+\left(y^+, Pr\right)$ depends on the dimensionless wall distance y^+ as well as on the Prandtl number Pr. This fact is of major importance, since only for $Pr \approx 1$, the thickness of the velocity and temperature boundary layer is approximately equal. In case of $Pr > 1$, the thermal boundary layer is thinner than the momentum boundary layer. Thus, special attention has to be paid to resolve the thermal boundary layer by an appropriate mesh resolution in the near-wall region. The dimensionless temperature θ^+ is zero directly at the wall whereas its profile within the *conduction layer* $\left(y^+ < 5\right)$ can be expressed through [28]:

$$\theta^+\left(y^+, Pr\right) = Pry^+. \tag{2.33}$$

The transition region $\left(5 < y^+ < 30\right)$ of the thermal boundary layer is similar to the momentum boundary layer, where turbulent heat transfer reaches the order of conduction. In the logarithmic region, which is dominated by turbulent heat transfer, the mean temperature profile can be described by [127]:

$$\theta^+\left(y^+, Pr\right) = \frac{1}{\kappa_\theta}\ln\left(y^+\right) + C_\theta^+(Pr), \tag{2.34}$$

where the constant κ_θ is usually set to $\kappa_\theta = 0.47$. For a smooth wall, the constant of integration C_θ^+ is now a function of the Prandtl number and can be approximated well by [53]:

$$C_\theta^+(Pr) = 13.7Pr^{2/3} - 7.5 \qquad (Pr > 0.5). \tag{2.35}$$

2.3 Mathematical modeling of turbulent flows

The governing equations for three-dimensional (incompressible) laminar, transitional and turbulent flows are the Navier-Stokes equations, derived in section 2.1. Since analytical solutions of these nonlinear PDEs are only known for very few simple laminar flow cases, mathematical turbulence modeling approaches and numerical simulation techniques are necessary to obtain an approximate solution of the flow fields (e.g., velocity $\mathbf{u}$) for a considered fluid dynamics problem. Based on the required computational effort and the resolved flow physics (simulation accuracy), the available treatments for turbulent flows can be roughly categorized as follows (see also Fig. 2.3):

- The fully resolved direct simulation (including the Kolmogorov scale) of the Navier-Stokes equation, also referred to as *direct numerical simulation* (DNS), is the most expensive technique even for moderate Reynolds numbers Re, since the required mesh resolution (number of grid cells N) for an accurate DNS scales with $N \propto Re^{9/4}$ [79]. Thus, the application of DNS is still restricted to the fundamental scientific research (e.g., detailed studies on combustion processes, development and adjustments of turbulence models).

- The calculation of practical high Reynolds-number flows is commonly performed by solving the steady or unsteady *Reynolds-averaged Navier-Stokes* equations (RANS) in combination with simplified turbulence models in which additional transport equations for different turbulence variables (e.g., turbulence kinetic energy k, dissipation rate ϵ) are solved to provide closure. The benefit of the RANS approach is the comparable low required CPU time, since only time-averaged flow fields are calculated and the turbulence is added through relatively easy to handle closure models. Unfortunately, the accuracy of RANS simulations often suffers from the high influence of the chosen turbulence model and its restrictions (e.g., assumption of isotropic turbulence based on the employed Boussinesq hypothesis [58]). Hence, a certain number of adjustments of the turbulence model to a particular flow situation are may required to obtain satisfactory results for practical applications [126]. More information and a rigorous discussion of the RANS approach can be found in [80, 166, 46].

- A technique with a level of generality in between DNS and RANS is the *large-eddy simulation* (LES) and can be seen as "poor mans DNS" [80]. In an LES, only the large eddies (or turbulent vortices) above a certain cut-off size are directly resolved on the underlying computational grid, while the small scales (*subgrid scales*) are considered through appropriate subgrid-scale (SGS) models. The fundamental idea behind this scale separation is that the smaller scales are much more isotropic and homogeneous than the large scales and are almost independent of the considered fluid dynamics problem, whereas the energy-containing large-scale vortices are strongly affected by the specific flow configuration. Due to the fact that not all turbulent scales have to be resolved on the numerical grid during an LES, only a fraction of the computational costs of an fully resolved DNS (typically of order 1%) is required [126]. Moreover, the influence of the SGS model on the accuracy of the numerical solution is drastically reduced in comparison to the RANS approach.

- Although LES represent a favorable opportunity for the prediction and analysis of complex turbulent flows (large-scale flow separation, turbulent flow control, combustion, etc.) at reduced computational costs, the application of LES to full configurations (e.g., ships, airplanes) at high Reynolds numbers still requires an immense computational effort. The main reason is the fine grid resolution necessary to resolve the boundary layer at the walls. To overcome this critical problem, one can apply *hybrid RANS/LES* approaches as the detached eddy simulation (DES), which combines the advantages of the LES and RANS approach. Thus, RANS is used in the vicinity of walls, whereas LES is employed for regions where large eddies occur, e.g., regions of flow separation. A comprehensive review of these methods for different practical applications can be found in [129, 124, 74].

Primarily the LES approach is used in this work to obtain approximate solutions of the carrier flow, since it allows the analysis of turbulent flows in which a representation of unsteady turbulent fluctuations is of high importance. This is the case for turbulent channel flows over dimpled surfaces as already reported by Turnow et al. [148, 149]. However, the application and assessment of hybrid RANS/LES models for these specific flow configurations was also part of this thesis and is presented in [147].

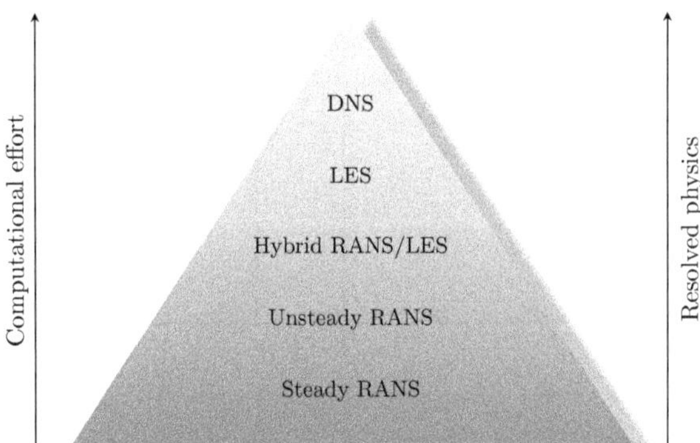

Fig. 2.3: Comparison of commonly used approaches for turbulent flows regarding the computational effort and resolved flow physics (adapted from [124]).

2.4 Large-eddy simulation

In this section, the spatial filtering or scale separation in turbulent flows and the derivation of the filtered transport equations (LES equations) is shortly discussed. Moreover, three different subgrid-scale model are presented, namely the "classical" Smagorinsky model and its dynamic version as well as the dynamic one equation eddy-viscosity model, which are extensively used in this work for closure (i.e., modeling of the subgrid-scale turbulence).

2.4.1 Spatial filtering

The separation of large and small scales in the LES approach proposed by Leonard [81] is achieved by introducing an appropriate low-pass filter G with the corresponding filter width $\overline{\Delta}$. Assuming the usual case of purely spatial and isotropic filtering $\left(\overline{\Delta}(\mathbf{x},t)=\overline{\Delta}\right)$ of any flow variable ϕ, the filtered variable $\overline{\phi}$ can be then formally expressed as convolution product by the filter kernel G:

$$\overline{\phi}(\mathbf{x},t) = G \star \phi(\mathbf{x}) = \int_{\Omega} G\left(\overline{\Delta}, \mathbf{x}-\mathbf{x}'\right)\phi\left(\mathbf{x}',t\right)\,\mathrm{d}^3\mathbf{x}', \tag{2.36}$$

where $G(\mathbf{x}-\mathbf{x}')$ is the rapidly decaying filter function and $\Omega \subset \mathbb{R}^3$ denotes the physical domain. In order to be able to manipulate the Navier-Stokes equations after applying a filter, it is essential that the filter G must verify the following fundamental properties [123]: conservation of constants, linearity and commutation with derivation. Typical low-pass filters used in the LES approach are the box filter, Gauss filter and the spectral (Fourier) filter, as shown in Fig. 2.4.

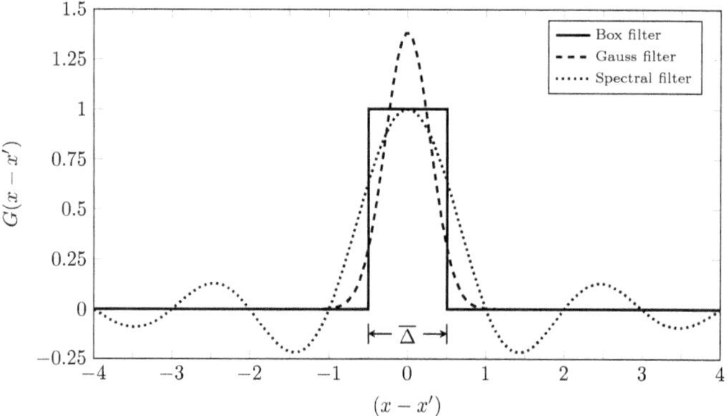

Fig. 2.4: Graphical illustration of filter function $G(x-x')$ in the one-dimensional physical space $(x-x')$ in case of box filter, Gauss filter and spectral (Fourier) filter for an equal filter width of $\overline{\Delta}=1$.

The box filtering is already implicitly realized in the framework of the finite volume method (FVM), which is also used in this work to solve the governing equations of fluid motion numerically. Thus, additional explicit filtering operations (see Lund [90]) are not necessary. A drawback of the box filter is its high sensitivity regarding the filter width $\overline{\Delta}$, especially for flows in complex geometries where a distortion and a strong grading of the computational grid is inevitable. This can causes commutation errors of second order $\mathcal{O}(\overline{\Delta}^2)$ [54], resulting in a reduction of the simulation accuracy. Additionally, it is difficult to separate numerical and filtering issues. However, the box filter is applied exclusively within this thesis, since the geometrical simplicity of all investigated flows allows a generation of block-structured computational grids of high quality (i.e., moderate distortion and grading of the mesh). Finally, it has to be noted that the filtered values will converge towards the instantaneous values as the filter width tends to be zero $\left(\overline{\Delta}\to 0\right)$.

Hence, the LES converges to the DNS results when the filter width is getting smaller than the smallest scales (i.e., Kolmogorov length scale η) of the turbulent flow.

2.4.2 Filtered transport equations

The filtered Navier-Stokes equations, often referred to as LES equations, and the filtered energy equation are derived by applying the filtering operation, Eq. (2.36), on the fundamental conservation equations for mass, momentum and energy, Eqs. (2.4), (2.6), and (2.14):

$$\nabla \cdot \overline{\mathbf{u}} = 0, \tag{2.37}$$

$$\frac{\partial \overline{\mathbf{u}}}{\partial t} + \nabla \cdot (\overline{\mathbf{u}}\,\overline{\mathbf{u}}) = -\nabla \overline{p} + \nu \nabla^2 \overline{\mathbf{u}} - \nabla \cdot \boldsymbol{\tau}_{SGS}, \tag{2.38}$$

$$\frac{\partial \overline{T}}{\partial t} + \nabla \cdot \left(\overline{\mathbf{u}}\overline{T}\right) = \nabla \cdot \left(a \nabla \overline{T}\right) - \nabla \cdot \mathbf{q}_{SGS}, \tag{2.39}$$

where the arising tensor $\boldsymbol{\tau}_{SGS}$ captures the non-resolved small scales and is dentoted as *subgrid-scale* (SGS) stress tensor:

$$\boldsymbol{\tau}_{SGS} = \overline{\mathbf{u}\mathbf{u}} - \overline{\mathbf{u}}\,\overline{\mathbf{u}}. \tag{2.40}$$

Analogously to the momentum equation, the scale separation of the temperature field results in an additional subgrid-scale heat flux $\mathbf{q}_{SGS}$. The closure of this term is accomplished in this work using a simple *gradient-diffusion hypothesis* (GDH), which assumes isotropic turbulence for simplicity:

$$\mathbf{q}_{SGS} = -a_t \nabla \overline{T} = -\frac{\nu_t}{Pr_t} \nabla \overline{T}, \tag{2.41}$$

where a_t is the turbulent thermal diffusivity, ν_t is the (subgrid-scale) eddy viscosity and Pr_t is the turbulent Prandtl number. The GDH is known for its inaccuracy in cases of highly anisotropic flows where the mean scalar gradient is not aligned with the scalar flux. However, the main advantage of the GDH closure is that no additional transport equations are required, making it relatively simple to implement numerically [24]. Following Leonard's decomposition[2] [81], a comprehensive analysis of different contributions of the SGS stress tensor, Eq. (2.40), can be conducted. Splitting of $\boldsymbol{\tau}_{SGS}$ by introducing the identity $\mathbf{u} = \overline{\mathbf{u}} + \mathbf{u}'$ leads to:

$$\boldsymbol{\tau}_{SGS} = \underbrace{\overline{\overline{\mathbf{u}}\,\overline{\mathbf{u}}} - \overline{\mathbf{u}}\,\overline{\mathbf{u}}}_{\mathbf{L}} + \underbrace{\overline{\mathbf{u}'\overline{\mathbf{u}}} + \overline{\overline{\mathbf{u}}\mathbf{u}'}}_{\mathbf{C}} + \underbrace{\overline{\mathbf{u}'\mathbf{u}'}}_{\mathbf{R}}, \tag{2.42}$$

where one can distinguishes between three different types of scale interactions which are designated with the three tensors $\mathbf{L}$, $\mathbf{C}$ and $\mathbf{R}$ [124]:

[2] It could be shown by Speziale [138] that two component stresses resulting from Leonard's decomposition are in fact not Galilean-invariant. An alternative, Galilean-invariant decomposition was later proposed by Germano [50].

- The *Leonard* stress tensor $\mathbf{L}$ which characterizes the fluctuations of the interactions between resolved scales themselves. This tensor can be directly calculated from the resolved flow variables.
- The *Cross* stress tensor $\mathbf{C}$ which characterizes the the cross interaction between the large (resolved) and small (subgrid scales of the flow.
- The *Reynolds* stress tensor $\mathbf{R}$ which accounts for the direct effect of the small flow scales on the resolved fields.

The SGS tensor is not closed since $\overline{\mathbf{uu}}$ cannot directly obtained from the filtered quantities $\overline{\mathbf{u}}$ alone. Thus, $\boldsymbol{\tau}_{SGS}$ must to be modeled using an appropriate SGS model which are further discussed in section 2.4.3.

It should be mentioned at this point that the filtered Navier-Stokes equations (2.38, 2.39) have a strictly similar form to the one of the Reynolds-averaged Navier-Stokes (RANS) equations, which corresponds to the special case of $\mathbf{L} = \mathbf{C} = \mathbf{0}$ in Eq. (2.42) and furthermore the application of Reynolds averaging instead of scale separation through low-pass filtering. This fact mainly motivates the development of hybrid RANS/LES approaches, as already mentioned in section 2.3.

2.4.3 Subgrid-scale modeling

This section provides a compact overview about the SGS models which have been applied to obtain the results presented in the chapters 6 and 7. A more rigorous discussion and analysis of various SGS models, their advantages and drawbacks can be found in [102, 80, 48, 124].

Smagorinsky model (SM)

One of the first and most-widely used SGS model is the Smagorinsky model, originally proposed by Smagorsinky in 1963 [130]. This "classical" eddy-viscosity model employs the Boussinesq hypothesis [58] for the computation of the anisotropic part of the SGS stress tensor $\boldsymbol{\tau}^a_{SGS}$:

$$\boldsymbol{\tau}^a_{SGS} = \boldsymbol{\tau}_{SGS} - \frac{1}{3}\,\mathrm{tr}\,(\boldsymbol{\tau}_{SGS})\mathbf{I} \approx -2\nu_t\overline{\mathbf{S}}, \tag{2.43}$$

where $\mathbf{I}$ is the identity tensor and $\overline{\mathbf{S}}$ is the large-scale strain-rate tensor defined as:

$$\overline{\mathbf{S}} = \frac{1}{2}\big(\boldsymbol{\nabla}\overline{\mathbf{u}} + \boldsymbol{\nabla}\overline{\mathbf{u}}^{\mathsf{T}}\big). \tag{2.44}$$

Assuming that the small scales are in equilibrium and dissipate entirely and instantaneously all the energy received from the resolved scales [30], the eddy viscosity ν_t is modeled through:

$$\nu_t = \left(C_S\overline{\Delta}\right)^2 |\overline{\mathbf{S}}|, \tag{2.45}$$

with the norm of the large-scale strain-rate tensor $|\overline{\mathbf{S}}| = \sqrt{\left(2\overline{\mathbf{S}} : \overline{\mathbf{S}}\right)}$. $\overline{\Delta}$ denotes the spatial filter width usually computed from the grid spacing Δ as:

$$\overline{\Delta} = (\Delta x \Delta y \Delta z)^{1/3}. \tag{2.46}$$

The model coefficient C_S, often referred to as *Smagorinsky constant*, has to be determined empirically and crucially depends on the investigated type of flow, the Reynolds number Re and the discretization schemes. Under the assumption that the cut-off frequency lies within a $\kappa^{-5/3}$ Kolmogorov cascade (see section 2.2), an approximation for C_S can be derived as follows [126]:

$$C_S = \frac{1}{\pi}\left(\frac{3C_K}{2}\right)^{-3/4} \approx 0.18, \tag{2.47}$$

where the Kolmogorov constant is assumed to be $C_K \approx 1.4$. Typical values of C_S in the literature range from 0.065 to 0.24 [157]. However, as reported by Deardorff [32] and later on by Moin and Kim [106], in practice the Smagorinsky constant has to reduced to values of $C_S \approx 0.1$ to sustain turbulence in channel flows. Remarkable advantages of the Smagorinsky model are the relatively straightforward application and its high numerical stability. On the other hand, the definition of a positive Smagorinsky constant ($C_S > 0$) results in a strictly positive dissipation rate ($\epsilon > 0$) which consequently does not allow the capturing of backscatter of turbulence kinetic energy. Moreover, the uniform specification of C_S for the entire computational domain suppresses the damping and complete vanishing of the eddy viscosity ν_t in the near-wall region and directly at the wall. This major drawback of the Smagorinsky model has to be corrected by additional damping functions as proposed by van Driest [150] or Piomelli et al. [114]. Moreover, the Smagorinsky model is not appropriate for simulating transition since it yields a positive eddy viscosity ($\nu_t \geq 0$) even in laminar flows.

Dynamic Smagorinsky model (DSM)

To overcome the disadvantageous of the Smagorinsky model (i.e., no reduction of the eddy viscosity ν_t in the near-wall region, no capturing of backscatter), Germano et al. [52] introduced a dynamic adjustment of the model coefficient C_S to the local flow conditions. Thus, an option to reduce the model contribution in the vicinity of walls or in laminar and transitional flow regions. This so-called *dynamic procedure* is based on an algebraic identity (also known as *Germano identity*) which relates the (known) subfilter stresses of a test filter to the corresponding (unclosed) subgrid-scales stresses of the grid filter [126]. Filtering of the LES equations (see section 2.4.2) by application of a coarser test filter $\widetilde{\overline{...}}$ ($\widetilde{\overline{\Delta}} > \overline{\Delta}$, e.g., half of the filter width of the (primary) grid filter [52]) leads to the subfilter-scale stresses $\mathbf{T}_{SGS}$ at the test-filter level:

$$\mathbf{T}_{SGS} = \widetilde{\overline{\mathbf{u}\mathbf{u}}} - \widetilde{\overline{\mathbf{u}}}\,\widetilde{\overline{\mathbf{u}}}. \tag{2.48}$$

Combining Eqs. (2.40) and (2.48) results in the Germano identity [52, 51], expressing the resolved turbulent stresses (or Leonard stress tensor, see section 2.4.2) as:

$$\mathbf{L} = \mathbf{T}_{SGS} - \widetilde{\boldsymbol{\tau}}_{SGS} = \left(\widetilde{\overline{\mathbf{u}\mathbf{u}}} - \widetilde{\overline{\mathbf{u}}}\,\widetilde{\overline{\mathbf{u}}}\right) - \left(\widetilde{\overline{\mathbf{u}\mathbf{u}}} - \widetilde{\overline{\mathbf{u}}\,\overline{\mathbf{u}}}\right) = \widetilde{\overline{\mathbf{u}}\,\overline{\mathbf{u}}} - \widetilde{\overline{\mathbf{u}}}\,\widetilde{\overline{\mathbf{u}}}, \tag{2.49}$$

By modeling the subgrid-scales stresses τ_{SGS} and subfilter stresses $\mathbf{T}_{SGS}$ using the closure suggested by Smagorinsky, Eqs. (2.43) and (2.45), one obtains:

$$\boldsymbol{\tau}_{SGS} - \frac{1}{3}\operatorname{tr}(\tau_{SGS})\mathbf{I} = -2\left(C_S\overline{\Delta}\right)^2|\overline{\mathbf{S}}|\overline{\mathbf{S}}, \tag{2.50}$$

$$\mathbf{T}_{SGS} - \frac{1}{3}\operatorname{tr}(T_{SGS})\mathbf{I} = -2\left(C_S\widetilde{\overline{\Delta}}\right)^2|\widetilde{\overline{\mathbf{S}}}|\widetilde{\overline{\mathbf{S}}}. \tag{2.51}$$

Employing the Germano identity, Eq. (2.49), yields the anisotropic part of the Leonard stress tensor:

$$\mathbf{L}^a = \mathbf{L} - \frac{1}{3}\operatorname{tr}(\mathbf{L})\mathbf{I} = 2\left(C_S\widetilde{\overline{\Delta}}\right)^2|\widetilde{\overline{\mathbf{S}}}|\widetilde{\overline{\mathbf{S}}} - 2\left(C_S\overline{\Delta}\right)^2\widetilde{|\overline{\mathbf{S}}|\overline{\mathbf{S}}} = 2C_{dyn}\mathbf{M}, \tag{2.52}$$

where $\mathbf{M} = \widetilde{\overline{\Delta}}^2|\widetilde{\overline{\mathbf{S}}}|\widetilde{\overline{\mathbf{S}}} - \overline{\Delta}^2\widetilde{|\overline{\mathbf{S}}|\overline{\mathbf{S}}}$ and C_{dyn} replaces the squared Smagorinsky constant C_S^2. It should be noted that for both filter levels the same spatially constant C_S was assumed, which cannot be realized in reality. However, the resulting error is negligible [48]. Applying the least-squares approach to minimize the quadratic error $\mathbf{L}^a - 2C_{dyn}\mathbf{M} = 0$, as suggested by Lilly [86], the model coefficient C_{dyn} can be determined from:

$$C_{dyn}(\mathbf{x}, t) = \frac{1}{2}\frac{(\mathbf{M} : \mathbf{L}^a)}{(\mathbf{M} : \mathbf{M})}. \tag{2.53}$$

The dynamic Smagorinsky coefficient $C_{dyn}(\mathbf{x}, t)$ determined through Eq. (2.53) is an instantaneous and local quantity varying in space $\mathbf{x}$ and time t. Furthermore, it can take negative values which indicates backscatter of kinetic energy. To avoid strong fluctuations of $C_{dyn}(\mathbf{x}, t)$ in space and therefore to increase the numerical stability, usually an additional spatial averaging in homogeneous direction is applied. For fully inhomogeneous flows in complex geometries, only averaging procedures in time (e.g., lowpass filtering) are applicable [20].

Dynamic one equation eddy-viscosity models (DOE)

One-equation models was introduced to avoid the simplifying assumption of equilibrium as used in the derivation of the Smagorinsky model (see 2.4.3). Non-equilibrium conditions commonly occur in free shear layers, separating and reattaching flows, boundary layers as well as wall-dominated flows like pipe and channels flows. Thus, the assumption of equilibrium is not valid for a large number of fluid dynamics problems. This issue can be overcome by adding a history effect, such as transport equations for one (or more) of the SGS turbulence characteristics, to the model [30]. Most of the one-equation models are also rely on the eddy-viscosity concept but additionally solve a transport equation for a SGS quantity, for instance the SGS kinetic energy $k_{SGS} = \mathrm{tr}(\boldsymbol{\tau}_{SGS})/2$, on which the eddy viscosity ν_t depends. A popular variant of an one-equation model is given by Yoshizawa and Horiuti [167, 168]:

$$\frac{\partial k_{SGS}}{\partial t} + \boldsymbol{\nabla} \cdot (\overline{\mathbf{u}}\, k_{SGS}) = (\nu + \nu_t)\nabla^2 k_{SGS} + \mathcal{P}_{SGS} - \epsilon_{SGS}. \tag{2.54}$$

The energy dissipation rate ϵ_{SGS}, which is assumed to be isotropic, and the production term of the unresolved kinetic energy $\mathcal{P}_{SGS}$ can found from:

$$\epsilon_{SGS} \approx C_e \frac{k_{SGS}^{3/2}}{\overline{\Delta}}, \tag{2.55}$$

$$\mathcal{P}_{SGS} = -\boldsymbol{\tau}_{SGS} : \overline{\mathbf{S}} \approx 2\nu_t \left(\overline{\mathbf{S}} : \overline{\mathbf{S}}\right), \tag{2.56}$$

where the eddy viscosity is given by:

$$\nu_t = C_k k_{SGS}^{1/2} \overline{\Delta}. \tag{2.57}$$

The model coefficients C_e and C_k can be either set to constant values or can be calculated dynamically, as described in section 2.4.3. Applications of the Germano identity, Eq. (2.49), yields:

$$C_k(\mathbf{x},\, t) = \frac{(\mathbf{M} : \mathbf{L}^a)}{(\mathbf{M} : \mathbf{M})}, \tag{2.58}$$

where $\mathbf{M} = -2\widetilde{\overline{\Delta} K^{1/2} \widetilde{\overline{\mathbf{S}}}}$ and $K = 1/2\left(\widetilde{|\overline{\mathbf{u}}|^2} - |\widetilde{\overline{\mathbf{u}}}|^2\right)$. The tensor $\mathbf{L}^a$ represents again the anisotropic part of the Leonard stress tensor $\mathbf{L}$. The model coefficient $C_e(\mathbf{x}, t)$ follows directly from [157]:

$$C_e(\mathbf{x}, t) = \frac{(\nu + \nu_t)\left(\widetilde{\overline{\mathbf{S}} : \overline{\mathbf{S}}} - \widetilde{\overline{\mathbf{S}}} : \widetilde{\overline{\mathbf{S}}}\right)}{\left(\widetilde{1/2 K^{3/2} \Delta^{-1}}\right)}. \tag{2.59}$$

Similar to the dynamic Smagorinsky model, the coefficients $C_k(\mathbf{x}, t)$ and $C_e(\mathbf{x}, t)$ significantly varying in space $\mathbf{x}$ and time t. To preserve the numerical stability and to avoid negative values of the eddy viscosity ($\nu_t \geq 0$) or rather backscattering of kinetic energy, an additional spatial averaging in homogeneous direction can be conducted.

3 Description of the dispersed phase

This chapter is mainly based on the work of Crowe et al. [25, 104] and Sommerfeld [135, 136], and provides a brief overview regarding the characterization of dispersed multiphase flows as well as commonly used numerical techniques to obtain approximate solutions for particles clouds. Moreover, the underlying equations of the applied Lagrangian particle tracking (LPT) approach is presented, used in this work to describe the dispersed phase in space $\mathbf{x}$ and time t. For convenience, the term dispersed phase will be used for the particulate phase, while continuous or carrier phase will be used for the carrier fluid. Parameters or quantities of the dispersed phase are denoted by subscript "p", whereas variables of the carrier flow are denoted by subscript "f".

3.1 Characterization of dispersed multiphase flows

Different integral properties are available to characterize dispersed multiphase flows. One of the most important (dimensionless) parameter is the so-called *volume fraction* of the dispersed phase, which is the volume occupied by particles in a unit volume:

$$\alpha_p = \frac{\sum_i N_i V_{p,i}}{V}, \tag{3.1}$$

where N_i is the number of all particles in the size fraction i, having the particle volume $V_{p,i} = \pi D_{p,i}^3/6$, and V is the considered volume. In case of non-spherical particles, the particle diameter $D_{p,i}$ is replaced by the volume equivalent diameter of a sphere. For some types of heavily deformed particles (e.g., fibers or droplets), this simplification is not suitable and an additional sphericity factor[1] should be introduced [135, 5]. Since the sum of the volume fraction of the dispersed and the continuous phase is unity, the volume fraction of the continuous phase (also referred to as *void fraction*) can be simply expressed as:

$$\alpha_f = 1 - \alpha_p. \tag{3.2}$$

The *bulk density* (or mass concentration) of the dispersed phase is defined as the mass of particles per unit volume:

$$\sigma_p = c_p = \frac{m_p}{V} = \alpha_p \rho_p, \tag{3.3}$$

[1]The spericity or form factor depends on the area or volume of the particle as seen by the flow.

with the particle density ρ_p, i.e., the particle weight divided by its volume V_p. Analogously, the bulk density of the continuous phase is:

$$\sigma_f = (1 - \alpha_p)\rho_f, \tag{3.4}$$

whereas the sum of both bulk densities is the so-called *mixture density*:

$$\rho_m = \sigma_f + \sigma_p = (1 - \alpha_p)\rho_f + \alpha_p\rho_p. \tag{3.5}$$

In many engineering applications and especially for gas-solid flow (e.g., pneumatic conveying), the *mass loading* is frequently used, defined as the total mass flux of the dispersed phase to that of the fluid phase:

$$\eta = \frac{\dot{m}_p}{\dot{m}_f} = \frac{\alpha_p\rho_p u_p}{(1 - \alpha_p)\rho_f u_f}, \tag{3.6}$$

where $\dot{m}_p$ and $\dot{m}_f$ is the mass flow rate of the dispersed and the continuous phase, respectively. The velocities for particles u_p and fluid u_f are averaged velocities in a considered cross-section. In addition to the volume fraction of the dispersed phase, the proximity of particles in a two-phase system is another important parameter for the classification of flow regimes and phase coupling mechanisms, and can be estimated from the *inter-particle spacing*, i.e., the distance between the centers of the particles. Assuming a homogeneous cubic arrangement of spherical particles, the dimensionless inter-particle spacing can be found from:

$$\frac{L}{D_p} = \left(\frac{\pi}{6\alpha_p}\right)^{1/3}. \tag{3.7}$$

Based on the introduced volume fraction of the dispersed phase α_p and the inter-particle spacing L/D, one can identify different regimes and momentum phase coupling mechanisms of dispersed two-phase flows, as presented in Fig. 3.1. Generally, it can be distinguished between *dilute* and *dense* two-phase flows. The assumption of a dilute two-phase system is valid for volume fractions up to $\alpha_p = 10^{-3}$ or an corresponding inter-particle spacing of $L/D \approx 8$, respectively. Furthermore, for dilute two-phase flows with a volume phase fraction of $\alpha_p < 10^{-6}$ (i.e., $L/D_P \approx 80$), the influence of the particle phase on the fluid flow can be neglected which is known as *one-way coupling*. With increasing volume fraction of the dispersed phase, the influence of the particle on the carrier flow has to be considered, which is also referred to as *two-way coupling*. In the case of dense two-phase systems (i.e., $\alpha_p > 10^{-3}$ and $L/D_p \lesssim 8$), inter-particle interactions (i.e., particle-particle collisions and fluid dynamic interactions between particles) become of importance. This regime is characterized by the so-called *four-way coupling*. An extended classification of dispersed two-phase flows was suggested by Elghobashi [41], taking the importance of phase coupling mechanisms and turbulence modulation (i.e., dissipation or enhancement of turbulence kinetic energy) into account (see also [67]). Elghobashi claimed that the dispersed phase in dilute one-way coupled two-phase flows has always an negligible effect on turbulence in the

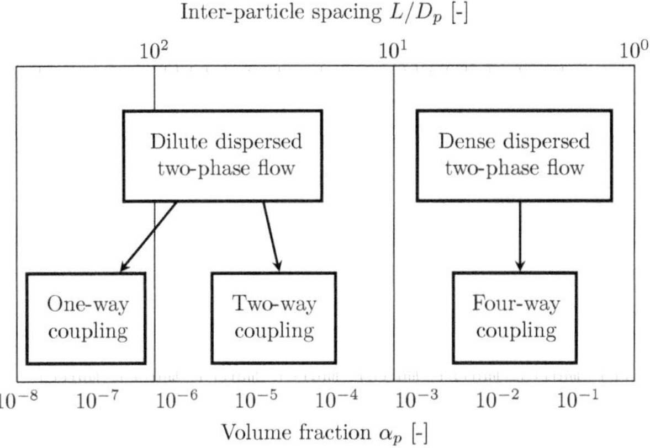

Fig. 3.1: Regimes and phase coupling mechanisms of dispersed two-phase flows based on the particle volume fraction α_p and inter-particle spacing L/D_p for a regular cubic arrangement of spherical particles (adapted from [136]).

carrier flow. In the case of two-way phase coupling, particles ether enhance the production ($\tau_p/\tau_\eta > 10^2$) or dissipation of turbulence ($\tau_p/\tau_\eta \leq 10^2$) in the fluid flow, depending on the Kolmogorov time scale τ_η (see section 2.2) and the so-called *particle relaxation time* or *momentum response time* τ_p. The momentum response time is of high importance for a further characterization of dispersed multiphase flows since it describes the capability of a particle to follow a sudden velocity change in the carrier flow. The response time can be derived from the general equation of motion, considering only the drag force:

$$m_p \frac{\mathrm{d}u_p}{\mathrm{d}t} = \frac{1}{2} C_D \frac{\pi D_p^2}{4} \rho_f \left(u_f - u_p\right) \left|u_f - u_p\right|, \tag{3.8}$$

where C_D is the drag coefficient, Eq. (3.17), u_p is the particle velocity and u_f is the carrier fluid velocity. Introducing the particle Reynolds number, Eq. (3.16), and dividing by the particle mass gives:

$$\frac{\mathrm{d}u_p}{\mathrm{d}t} = \frac{18\mu_f}{\rho_p D_p^2} \frac{C_D Re_p}{24} \left(u_f - u_p\right), \tag{3.9}$$

where μ_f is the dynamic viscosity of the carrier fluid. In the Stokes regime (low Reynolds numbers), the term $C_D Re_p/24$ approaches to unity, whereas the first factor has the dimension of a reciprocal time, the momentum response time:

$$\tau_p = \frac{\rho_p D_p^2}{18\mu_f}. \tag{3.10}$$

Hence, the equation of motion can be rewritten as:

$$\frac{\mathrm{d}u_p}{\mathrm{d}t} = \frac{1}{\tau_p}\big(u_f - u_p\big). \tag{3.11}$$

For a simple jump of the fluid velocity from zero to a constant u_f and an initial particle velocity of zero, the solution of Eq. (3.11) is:

$$u_p = u_f[1 - \exp\left(-t/\tau_p\right)]. \tag{3.12}$$

Therefore, the particle response time is the time required for a particle at rest, to reach 63.2% of the freestream flow velocity u_f. This characteristic time has been used widely in the field of multiphase flows and can be further extended to the (turbulence) *Stokes number*, which is the ratio of the particle response time τ_p to a characteristic time scale of the turbulence τ_f within the carrier flow:

$$St = \frac{\tau_p}{\tau_f}. \tag{3.13}$$

Accordingly, if the Stokes number tends to zero (e.g., for tiny particles), the particles follow changes of the carrier flow immediately and are almost homogeneously distributed. For dispersed multiphase flows in which the Stokes number becomes very large, the behavior of the particles is highly uncorrelated with the flow characteristics, as illustrated in Fig. 3.2.

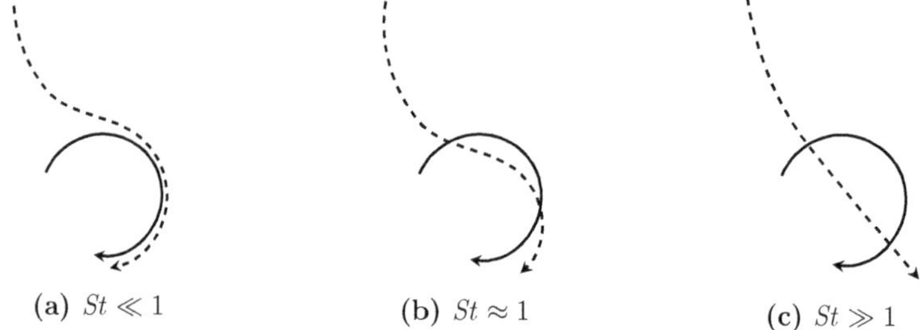

(a) $St \ll 1$ (b) $St \approx 1$ (c) $St \gg 1$

Fig. 3.2: Influence of a turbulent eddy (solid line) on the trajectory of a particle (dashed line) for different Stokes number limits (adapted from [5]).

3.2 Numerical modeling of dispersed multiphase flows

The choice of a suitable approach for the numerical simulation of the dispersed phase (or particle cloud) crucially depends on the dense or dilute character of the two-phase flow. The basic methods for modeling particle clouds can be categorized into *Lagrangian tracking* or *Eulerian modeling* approaches. In dilute flows, the time between particle collisions is much larger than the momentum response time which means that the particle-fluid interaction is mainly responsible for the particle motion instead of inter-particle interactions. Thus, information regarding the dispersed phase (e.g., particle velocity, size) travels along specific particle trajectories. In this case, only the Lagrangian approach can be employed for numerical simulations. In dense flows, particle collisions allow the spread of information in all spatial directions and the particle cloud can be modeled as continuous phase using the Eulerian approach. The basic approaches for modeling particle clouds are shown in Fig. 3.3 and are shortly described below [25, 5, 104]:

- In the *Discrete Element Method* (DEM), the motion of each particle is determined considering the fluid dynamic forces, the contact forces and moments due to the neighboring particles. Contact forces and moments are captured by a so-called soft-sphere collision model (see section 3.6.1), originally proposed by Candall and Struck [27] and further developed by Tsuji et al. [145]. This approach has found numerous applications in granular flows, fluidized beds and conveying systems and is particularly applicable to contact-dominated flows. The DEM allows a very detailed description of the dispersed phase, but typically a large number of discrete elements are required to model practical systems which results in enormous computational costs.

- In the *Discrete Parcel Method* (DPM), a group (parcel) of particles is identified and tracked through the flow field. All particles in the parcel are assumed to have the same properties (velocity, size, etc.) so the group is represented by one (representative) computational particle. Basically, two different approaches to the DPM exist. One approach applies to dilute and near-dilute two-phase flows in which the fluid dynamic forces (i.e., drag and lift) acting on the particle are responsible for the motion of the particles. Initially introduced as Particle-Source-In-Cell (PSI-Cell) method by Crowe et al. [26], the extension of this approach by additional sub-models (e.g., turbulence dispersion, particle collision or agglomeration) is relatively straightforward. The second DPM approach is known as the Multiphase Particle-In-Cell method (MP-PIC), originally proposed by Andrews and O'Rourke [6], in which the particle motion equations include the effect of particle-particle interaction by incorporating the gradient of the solids stress tensor as a force on the computational particle.

- In the *Two-Fluid Model* (TFM), the particles cloud is described as an additional continuous fluid phase. The differential governing equations for the conservation of mass, momentum and energy are developed at a point in the cloud. These equations are discretized into algebraic equations at the computational nodes in the field and solved using the same numerical procedure as used for the carrier flow.

In this work, a Lagrangian algorithm based on the *point-particle tracking* [39] and the DPM is used to describe the dispersed phase (i.e., foulant particles). Hence, the tracked particles (or parcels) are "point" particles which occupy no physical volume. In order to determine the fluid dynamic forces acting on the particle cloud, the carrier flow velocity is provided at every point in the flow through appropriate interpolation methods. Additionally, different sub-models for particle deposition, particle resuspension and phase conversion are implemented into the Lagrangian framework (see Kasper et al. [67, 66, 68]), which are also described in chapter 5.

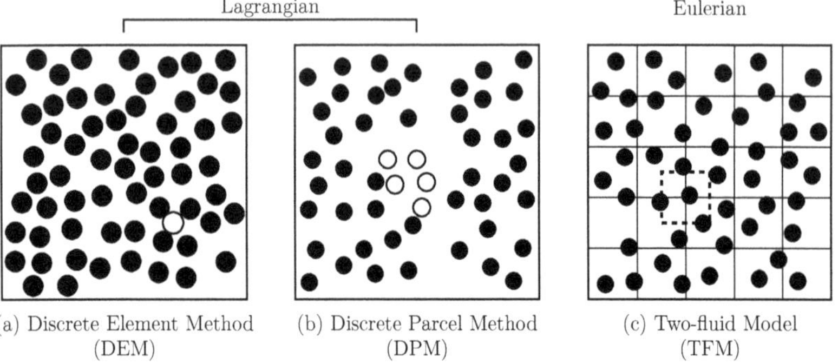

(a) Discrete Element Method (DEM) (b) Discrete Parcel Method (DPM) (c) Two-fluid Model (TFM)

Fig. 3.3: Basic approaches for modeling particle clouds (adapted from [25]).

3.3 Forces on particles

In the Lagrangian framework, the motion of particles (i.e., change of particle location, linear and angular components the particle velocity) is calculated by solving a set of ordinary differential equations along the particle trajectories. According to Newton's second law of motion, this requires the consideration of all relevant forces acting on the particle. The so-called BBO-equation, based on the fundamental work of Basset [11], Boussinesq [18] and Oseen [113], describes the motion of small particles in a quiescent viscous fluid (i.e., for small particle Reynolds numbers $Re_p < 0.5$, also referred to as Stokes regime). Based on the BBO-equation, Maxey and Riley [101] proposed the equation of motion for small particles in non-uniform Stokes flow. Unfortunately, this equation cannot be directly applied to particle motions in turbulent carrier flows. Therefore, the LPT approach used in this work allows a rigorous extension of the general BBO-equation for higher particle Reynolds numbers. Since analytical solutions for the majority of the particle forces are only available for small particle Reynolds numbers (i.e., Stokes regime) [25], this extension is based on empirical correlations derived from experiments. Considering linear motion of rigid, spherical particles and neglecting any heat and mass transfer, the calculation of the particle trajectories requires the solution of the following two ordinary differential equations:

$$\frac{\mathrm{d}\mathbf{x}_p}{\mathrm{d}t} = \mathbf{u}_p, \tag{3.14}$$

$$m_p \frac{\mathrm{d}\mathbf{u}_p}{\mathrm{d}t} = \sum \mathbf{F}_\mathrm{i}, \tag{3.15}$$

where $\mathbf{x}_p$ is the particle location and m_p is the mass of the particle. The term $\mathbf{F}_i$ in Eq. (3.15) accounts for all relevant forces acting on the particle (e.g., drag and gravity forces) and their determination will be shortly discussed in the following sections.

3.3.1 Drag force

In most particle-laden flows the drag force is dominating for the particle motion [104] and is expressed in terms of a drag coefficient C_D, which allows the calculation of the drag force not only for small particle Reynolds numbers ($Re_p < 0.5$) where viscous effects are dominating and no separation is observed, but also for the transition region (i.e., $0.5 < Re_p < 1{,}000$) and fully turbulent region or Newton regime (above $Re_p \approx 1{,}000$). The implemented drag model is based on the particle Reynolds number, which is defined as:

$$Re_p = \frac{\rho_f D_p \left|\mathbf{u}_f - \mathbf{u}_p\right|}{\mu_f}, \tag{3.16}$$

with the density ρ_f and the dynamic viscosity μ_f of the fluid or continuous phase, the particle diameter D_p and the magnitude of the relative slip velocity $\left|\mathbf{u}_f - \mathbf{u}_p\right|$. The drag coefficient is determined in this work using the following drag model based on the correlation proposed by Putnam [118]:

$$C_D = \begin{cases} \frac{24}{Re_p}\left(1 + \frac{1}{6} Re_p^{2/3}\right) & \text{if} \quad Re_p \leq 1{,}000 \\ 0.424 & \text{if} \quad Re_p > 1{,}000, \end{cases} \tag{3.17}$$

which is suitable to high Reynolds numbers ($Re_p < 1{,}000$) and ensures the correct limiting behavior within the Newton regime. Based on the drag coefficient, the general equation of particle motion, Eq. (3.8), is used to evaluate the corresponding drag coefficient for a spherical particle[2] as follows:

$$\mathbf{F_D} = C_D \frac{\pi D_p^2}{8} \rho_f \left(\mathbf{u}_f - \mathbf{u}_p\right)\left|\mathbf{u}_f - \mathbf{u}_p\right|. \tag{3.18}$$

3.3.2 Other forces

Depending on the particular flow conditions, other forces acting on the particles become more and more important. Especially in dispersed two phase flows with a density ratio ρ_p/ρ_f close to unity (i.e., liquid-solid flows), the consideration of additional forces as the pressure gradient force, gravitational and buoyancy force as well as added mass force is of importance [25]. In this work, the employed particle-tracking algorithm is capable to consider most of the forces, which commonly arises in dispersed two-phase flows, as described by Kasper et al. [67].

[2]The particle shape strongly influences the drag force. In case of non-spherical particles, the shape effect is considered through extended correlations of the drag coefficient which includes an additional sphericity or form factor (see e.g., Haider and Levenspiel [55]).

3.4 Phase coupling

Phase coupling is an important concept in dispersed two-phase flows and was already introduced in section 3.1. Here is shown how the different mechanisms of phase coupling are realized in this work. It should be mentioned at this point that, although coupling can take place through mass, momentum and energy transfer between the phase, only momentum coupling is considered in the following numerical studies. Fig. 3.4 shows a schematic diagram of the different momentum coupling mechanisms occurring in dispersed two-phase flows.

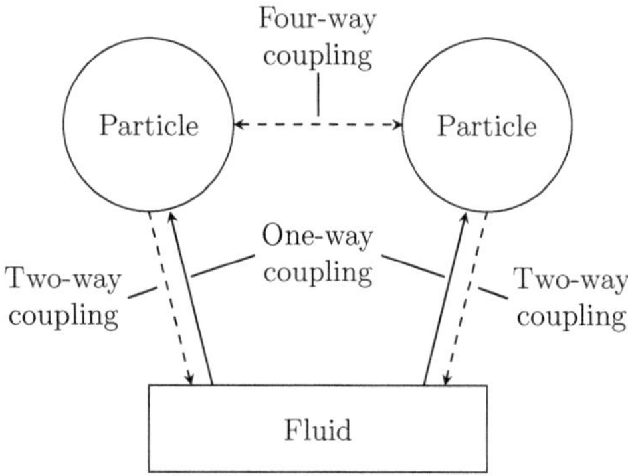

Fig. 3.4: Schematic diagram of momentum coupling in dispersed two-phase flows (adapted from [104]).

The dispersed two-phase flow is one-way coupled, if the carrier flow affects the dispersed phase while there is no reverse effect. This is implemented by solving the filtered momentum equation, Eq. (2.39), without any additional source term contribution ($\overline{\mathbf{S}}_M = \mathbf{0}$). If there is a mutual effect between the continuous and dispersed phase, then the flow system is two-way coupled. In this case, the source term contribution $\overline{\mathbf{S}}_M$ to the filtered momentum equation, Eq. (2.39), of all particles which have been in the computational grid cell k (with the corresponding volume V_k) during the Eulerian time step $\mathrm{d}t$ is written as [4]:

$$\overline{\mathbf{S}}_M = -\frac{1}{V_k\,\mathrm{d}t}\sum_p m_p\Big((\overline{\mathbf{u}}_p)_{t,out} - (\overline{\mathbf{u}}_p)_{t,in}\Big), \tag{3.19}$$

where m_p is the particle mass, $(\overline{\mathbf{u}}_p)_{t,in}$ and $(\overline{\mathbf{u}}_p)_{t,out}$ are the particle velocities (based on the filtered velocity field), when entering and leaving grid cell k. The applied LPT algorithm (see section 4.5) is able to compute the portion of time step $\mathrm{d}t$ particles spent in each computational cell. In order to account for four-way coupling, additional inter-particle collision models have to be used which are presented in the following section 3.6.1.

3.5 Particle dispersion

The applied particle-tracking approach requires the instantaneous fluid velocity $\mathbf{u}_f$ along the particle trajectory. In LES of turbulent dispersed flows, modeling inaccuracies are incurred since LES provides only an approximation of the instantaneous velocity field (i.e. filtered velocity). These inaccuracies can affect the estimation of the forces acting on the particles, when the filtered velocity $\overline{\mathbf{u}}_f$ without consideration of the non-resolved subgrid velocity fluctuations is supplied to the fundamental equations of particle motion. As a result, particle trajectories in LES fields progressively diverge from the actual particle trajectories and the flow fields seen by the particles become less and less correlated (i.e. forces acting on the particles are evaluated at increasingly different locations). A comprehensive discussion and review of existing models and strategies to solve this issue is presented by Marchioli [96]. In this work, the contribution of the non-resolved subgrid velocity fluctuations on the particle motions is modeled using a stochastic approach proposed by Pozorski and Apte [117], based on the assumption of isotropic turbulence. Accordingly, the standard deviation σ can is described through:

$$\sigma = \sqrt{\frac{2}{3} k_{SGS}}; \; \sigma = \sqrt{\overline{u'^2}} = \sqrt{\overline{v'^2}} = \sqrt{\overline{w'^2}}. \tag{3.20}$$

Multiplication of the standard deviation σ with a random vector $\boldsymbol{\zeta}$ following a Gaussian distribution with zero mean and unit variance allows one to estimate the resulting fluid velocity by superposition of the filtered velocity $\overline{\mathbf{u}}_f$ and the modeled SGS velocity fluctuations $\mathbf{u}'_f$:

$$\mathbf{u}_f = \overline{\mathbf{u}}_f + \mathbf{u}'_f \text{ with } \mathbf{u}'_f = \boldsymbol{\zeta}\sigma. \tag{3.21}$$

At each particle position $\mathbf{x}_p$, the instantaneous fluid velocity $\mathbf{u}_f$ is determined through Eq. (3.21) and is considered in the equations of particle motion, Eqs. (3.14) and (3.15). The consideration of unresolved velocity fluctuations is of importance especially for fine particles. In this case, the corresponding particle relaxation time is of the same order as the smallest fluid time scales [19].

3.6 Particle interactions

The particle-laden flows presented in this work can be categorized as wall-dominant, dilute or near-dilute two-phase flows ($\alpha_p < 10^{-3}$). In this case, particle motions are mainly controlled by particle-wall interactions rather than by particle-particle collisions. However, this sections gives an overview regarding modeling strategies for both collision scenarios, inter-particle collisions and particle-wall interactions.

3.6.1 Particle-particle collisions

In general, inter-particle (or particle-particle) collision is negligible in dilute and near-dilute particle-laden flows. As the particle concentration becomes higher, particles collide with each other and the resulting loss of particle kinetic energy, due to the inelastic property of the particle material, has to be taken into account. Two types of collision models are widely employed to account for particle-particle collisions, namely the *hard sphere* model and the *soft sphere* model.

The hard sphere model describes the dynamics of individual, binary particle-particle collisions in terms of conservation of momentum and energy. Assuming a collision of rigid spherical particles (i.e., ideal collision with full conservation of momentum), the translational momentum transferred during a collision can be quantified as [5]:

$$m_{p,1}\left(\mathbf{u}_{p,1} - \mathbf{u}'_{p,1}\right) = m_{p,2}\left(\mathbf{u}_{p,2} - \mathbf{u}'_{p,2}\right) = \mathbf{J}, \tag{3.22}$$

where $m_{p,1}$ and $m_{p,2}$ represents the mass of particle 1 and particle 2, $\mathbf{J}$ is the unknown impulsive force exerted on particle 1 (which also acts on particle 2 as the reaction force). $\mathbf{u}_{p,1}$ and $\mathbf{u}_{p,2}$ is the particle velocity before collision, whereas $\mathbf{u}'_{p,1}$ and $\mathbf{u}'_{p,2}$ are the corresponding particle velocities after collision. In Eq. (3.22), the particle mass, velocities before collision, and positions before collisions are given. Thus, the unknown variables are the impulsive force $\mathbf{J}$ and the post-collision velocities $\mathbf{u}'_{p,1}$ and $\mathbf{u}'_{p,2}$. Introducing the restitution coefficient e as:

$$e = \frac{\mathbf{u}'_{p,2} - \mathbf{u}'_{p,1}}{\mathbf{u}_{p,1} - \mathbf{u}_{p,2}}, \tag{3.23}$$

Eq. (3.22) can be solved analytically under different assumptions. The solution (i.e., post-collision velocities $\mathbf{u}'_{p,1}$ and $\mathbf{u}'_{p,2}$) is described in [25]. Major advantages of the hard sphere model are the easy handling and the low computational effort, but it is applicable only to binary collisions[3].

In the soft sphere model, the local deformation of the particles during collision is determined. This deformation is due to reversible deformation of the particle and is related to the linear overlap distance δ of the particles during collision. The force resulting from the deformation is typically written in terms of a so-called *spring-slider-dash-pot* model, as shown in Fig. 3.5. The normal component of the contact force, $\mathbf{F}_{n,ij}$, acting on particle i is given by the sum of the forces due to the spring and dash-pot [25]:

$$\mathbf{F}_{n,ij} = (-k_n \delta_n^\alpha - \eta_{n,j} \mathbf{G} \cdot \mathbf{n})\mathbf{n}, \tag{3.24}$$

where δ_n is the displacement (or overlap distance) of particle caused by the normal force, k_n is the stiffness coefficient of the spring in normal direction, $\eta_{n,j}$ is the damping coefficient of the dash-pot in normal direction, $\mathbf{G}$ is the velocity vector of particle i relative to particle j ($\mathbf{G} = \mathbf{u}_{p,i} - \mathbf{u}_{p,j}$), and $\mathbf{n}$ is the unit vector in the direction of the line from the center of particle i to that of particle j. The exponent α in Eq. (3.24) is $\alpha = 1$ for

[3] For a dispersed particulate phase, it is sufficient to consider binary collisions [25].

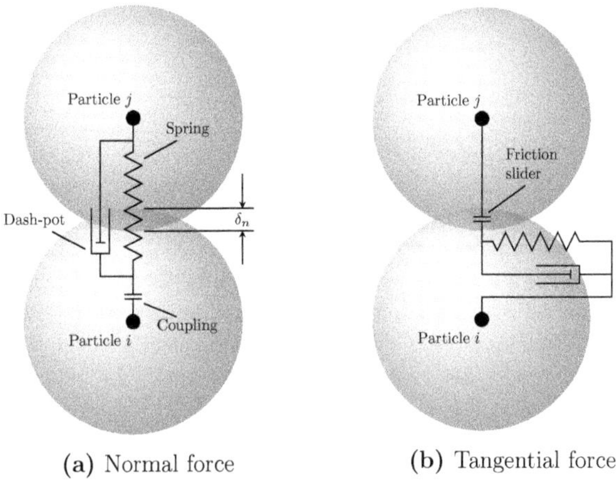

(a) Normal force (b) Tangential force

Fig. 3.5: Illustration of the spring-slider-dashpot concept in context of the soft sphere model for normal force and tangential force.

simple two-dimensional collisions (i.e., particles are assumed to be two-dimensional disks) and $\alpha = 3/2$ (according to the *Hertzian contact theory*) in the case of three-dimensional spheres [5, 104]. The tangential component of the contact force, $\mathbf{F}_{t,ij}$, can be expressed as:

$$\mathbf{F}_{t,ij} = -k_t\delta_t - \eta_{t,j}\mathbf{G}_{ct}, \tag{3.25}$$

where k_t and $\eta_{t,j}$ are, respectively, the stiffness and damping coefficient in the tangential direction, and $\mathbf{G}_{ct}$ is the slip velocity if the contact point. Assuming only translational motions of the particles, the slip velocity is given as:

$$\mathbf{G}_{ct} = \mathbf{G} - (\mathbf{G} \cdot \mathbf{n})\mathbf{n}. \tag{3.26}$$

If the condition of sliding is satisfied:

$$|\mathbf{F}_{t,ij}| > \mu|\mathbf{F}_{n,ij}|, \tag{3.27}$$

particle i starts to slide and the tangential force is given by the Coulomb-type friction law, Eq. (3.28), instead of Eq. (3.25):

$$\mathbf{F}_{t,ij} = -\mu|\mathbf{F}_{n,ij}|\mathbf{t}, , \tag{3.28}$$

where μ is the friction coefficient and $\mathbf{t} = \mathbf{G}_{ct}/|\mathbf{G}_{ct}|$ is the corresponding unit vector. For high volume fractions of the dispersed phase, multiple particles are in contact with particle

i at the same time. Therefore, the total contact force acting on particle i is obtained through the sum of the normal and tangential component of the contact force with respect to j:

$$\mathbf{F}_i = \sum_j (\mathbf{F}_{n,ij} + \mathbf{F}_{t,ij}). \tag{3.29}$$

The determination of the remaining unknown parameters, namely the stiffness k, damping coefficient η, and friction coefficient f can be found in [145, 25]. The main benefit of the soft sphere model is its applicability even for contact-dominated, dense two-phase flows, since no restrictions regarding binary particle-particle collisions exist. The major drawback of the soft sphere model is the much longer computation time (or high computational effort) compared to the hard sphere model [104]. Consequently, the application of this collision model is only applicable to particle clouds consisting of a moderate number of particles.

3.6.2 Particle-wall interaction

The effect of a wall is to retard the particle motion in normal and tangential direction. Similar to inter-particle collisions, two modeling approaches for particle-wall interactions are commonly used. Assuming a contact partner of infinite size, the soft sphere model discussed in section 3.6.1 can be directly applied to solve for particle-wall collisions. To overcome the disadvantages of the soft sphere model (e.g., high computational effort), a hard sphere model is used in this work to simulate the wall rebound of the particles, which allows a very efficient calculation of (binary) particle-wall interactions even for a large number of particles. Being u_p and u_p' the velocities of the considered particle before and after the wall collision and $\mathbf{n}$ and $\mathbf{t}$ the unit vector normal and tangential to the wall, the particle velocities can be written as follows:

$$\mathbf{u}_p = u_p^n \cdot \mathbf{n} + u_p^t \cdot \mathbf{t}, \tag{3.30}$$

$$\mathbf{u}_p' = u_p'^n \cdot \mathbf{n} + u_p'^t \cdot \mathbf{t}. \tag{3.31}$$

The normal component of the particle velocity after the wall collision is calculated as:

$$u_p'^n = -e \cdot u_p^n, \tag{3.32}$$

where e is the coefficient of normal restitution to the wall. The tangential component of the velocity will decrease after the collision with the wall an is evaluated as:

$$u_p'^t = (1 - \mu_w) \cdot u_p^t, \tag{3.33}$$

where μ_w is the coefficient of friction to the wall. Setting $e = 1$ and $\mu_w = 0$ corresponds to a perfect elastic collision (i.e., no dissipation of kinetic energy) without tangential friction at the wall.

4 Numerical methodology

This chapter is devoted to a short introduction into the cell-centered[1] finite volume (FV) method which is used in the present work to solve the conservation equations (2.4, 2.10 and 2.14) numerically. Among other existing discretization approaches (e.g., finite difference (FD) method, finite element (FE) method or spectral methods), the FV method is probably the most popular approach for engineering applications. Major advantages of this method are the compatibility with any kind of unstructured type of grid, which is favorable for complex geometries, and the physical meaning of each discretized term. Additionally, the FV method is conservative by construction and is relatively easy to program. In-depth discussions and further details on the FV method can be found in [64, 151, 59, 46].

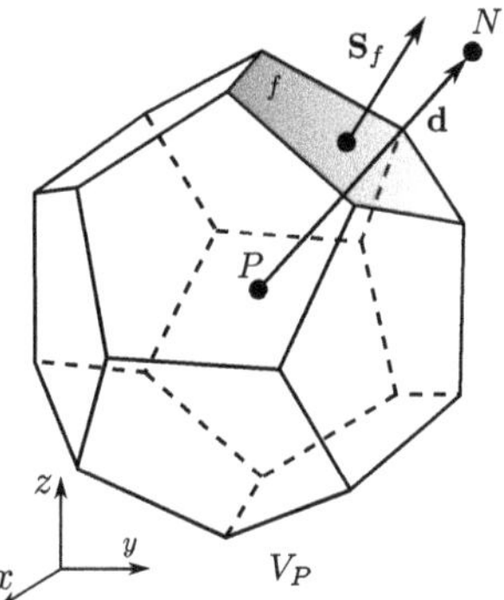

Fig. 4.1: Arbitrarily-shaped CV (grid cell) used in the FV method (adapted from [156]).

The basic concept of the FV method is to subdivide the solution domain into a finite number of contiguous control volumes (CVs) and to apply the conservation equations to each of them. Fig. 4.1 shows a typical arbitrarily-shaped control volume used for the spatial discretization, where P and N are the computational points - or cell centers - of the *owner* and *neighbor* CV. The computational point P is located at the centroid of the CV:

$$\int_{V_P} (\mathbf{x} - \mathbf{x}_P) \, \mathrm{d}V = 0, \tag{4.1}$$

[1]All variables are calculated and stored at the centroid of each control volume (grid cell).

with the volume V_P of the considered CV and the position vector $\mathbf{x}$. Two adjacent control volumes are connected via cell face f with the related face area vector $\mathbf{S}_f$, which is defined to be perpendicular to f and points outward of the considered CV into the neighboring CV. The magnitude of $\mathbf{S}_f$ equals the surface area of cell face f. Similar to Eq. (4.1), the mid point of this face matches the surface centroid $\mathbf{x}_f$ of f:

$$\int_S \left(\mathbf{x} - \mathbf{x}_f\right) \mathrm{d}S = 0. \tag{4.2}$$

The vector $\mathbf{d}$ represents the connection of the cell-center points P and N. If $\mathbf{d}$ is perpendicular to f the grid is called *orthogonal*, which is preferential for numerical simulations. The discretization procedure on non-orthogonal meshes needs special attention and will be further discussed in section 4.1. In the case that the point of intersection between $\mathbf{d}$ and f does not match the location of the surface centroid of f, the grid is referred to as *skew*.

4.1 Spatial discretization

The following explanation of the spatial discretization procedure is based on a generic convection-diffusion equation which describes the transport and conservation of an arbitrary physical quantity ϕ (e.g., mass, velocity or momentum). The standard form of this expression[2] reads:

$$\underbrace{\frac{\partial \rho\phi}{\partial t}}_{\text{temporal derivative}} + \underbrace{\nabla \cdot (\rho \mathbf{u} \phi)}_{\text{convection term}} - \underbrace{\nabla \cdot \left(\rho \Gamma_\phi \nabla \phi\right)}_{\text{diffusion term}} = \underbrace{S_\phi(\phi)}_{\text{source term}} , \tag{4.3}$$

where Γ_ϕ is the diffusion coefficient and $S_\phi(\phi)$ is the source term. The left-hand side of this equation accounts for the temporal change of ϕ as well as for the convection and diffusion of ϕ, whereas all remaining terms are summarized in source term $S_\phi(\phi)$ on the right-hand side. Within the framework of the finite volume method, Eq. (4.5) is integrated over control volume V_P (around point P) and the time t:

$$\int_t^{t+\Delta t} \left[\frac{\partial}{\partial x} \int_{V_P} \rho\phi \mathrm{d}V + \int_{V_P} \nabla \cdot (\rho \mathbf{u} \phi) \mathrm{d}V - \int_{V_P} \nabla \cdot \left(\rho \Gamma_\phi \nabla \phi\right) \mathrm{d}V\right] \mathrm{d}t \tag{4.4}$$

$$= \int_t^{t+\Delta t} \left(\int_{V_P} S_\phi(\phi) \mathrm{d}V\right) \mathrm{d}t. \tag{4.5}$$

---

[2] This generic transport equation is of second order. To ensure a high numerical accuracy, the order of the discretization has to be equal to or higher than the order of the equation that is being discretized [64, 46].

The volume integrals can be rigorously converted into surface integrals by application of the Gauss' theorem [8]:

$$\int_V \nabla \circ \phi \mathrm{d}V = \oint_{S_V} \mathbf{n} \circ \phi \mathrm{d}S = \oint_{S_V} \mathrm{d}\mathbf{S} \circ \phi, \tag{4.6}$$

where S_V is the closed surface of control volume V and $\mathrm{d}\mathbf{S}$ is the surface-normal vector (points out of the control volume) of an infinitesimally small surface element. The operator $\circ$ represents any tensor product, i.e., for the divergence, gradient and rotation of quantity ϕ. Since each control volume is bounded by a set of flat cell faces f, Eq. (4.6) can be rewritten as:

$$\int_V \nabla \circ \phi \mathrm{d}V = \oint_{S_V} \mathrm{d}\mathbf{S} \circ \phi = \sum_f \left(\int_f \mathrm{d}\mathbf{S} \circ \phi \right). \tag{4.7}$$

The surface integral in Eq. (4.7) can now be approximated with second-order accuracy as a product of the integrand at the cell-face center and the cell-face area (midpoint-rule approximation) [46]:

$$\int_f \mathrm{d}\mathbf{S} \circ \phi \approx \mathbf{S} \circ \phi_f, \tag{4.8}$$

with the outward-pointing face area vector $\mathbf{S}$ of the cell face f used for integration. According to Fig. 4.1, the face area vector S_f is "owned" by P and points outward V_P. Hence, S_f points inwards for all "neighboring" cell faces which needs to be taken into account in Eq. (4.8). The sum over the cell faces is therefore split[3] into sums over owned and neighboring faces [64]:

$$\sum_f = \mathbf{S} \circ \phi_f = \sum_{owner} \mathbf{S}_f \circ \phi_f - \sum_{neighbor} \mathbf{S}_f \circ \phi_f. \tag{4.9}$$

According to Eqs. (4.8) and (4.9), the spatial discretization requires the interpolation of the cell-face value ϕ_f (in the middle of face f) through the corresponding cell-center values ϕ_P and ϕ_N using appropriate numerical schemes, which will be discussed within the following section.

Convective term

The spatial discretization of the convective term can be achieved using Eqs. (4.7) and (4.8):

$$\int_{V_P} \nabla \cdot (\rho \mathbf{u} \phi) \mathrm{d}V = \int_S \mathrm{d}\mathbf{S} \cdot (\rho \mathbf{u} \phi) \approx \sum_f \mathbf{S} \cdot (\rho \mathbf{u})_f \phi_f = \sum_f F \phi_f, \tag{4.10}$$

[3] In the rest of this work, this split is automatically assumed.

where $F = \mathbf{S} \cdot (\rho\mathbf{u})_f$ represents the convection mass flux through the cell face. A separate discussion of the calculation of these cell face fluxes is presented in section 4.3. For now it can be assumed that F is calculated from the interpolated values of ρ and $\mathbf{u}$. The needed interpolations are performed by means of so-called face interpolation or convection difference schemes. A comprehensive overview of differencing schemes and detailed investigations regarding the accuracy and boundedness can be found in [64, 59, 151, 162, 46]. Among various existing face interpolation strategies, one of the most intuitive way is to approximate ϕ_f by its nearest upstream/downstream cell-center value ϕ_P or ϕ_N (depending on the flow direction). This approximation is called *upwind differencing* (UD) and ϕ_f follows from:

$$\phi_f = \begin{cases} \phi_f = \phi_P & \text{if} \quad F \geq 0, \\ \phi_f = \phi_N & \text{if} \quad F < 0. \end{cases} \tag{4.11}$$

The UD is the only approximation that unconditionally satisfies the boundedness criterion (i.e., the UD scheme never produces oscillatory solutions), but it is only of first-order accuracy and introduces additional numerical diffusion. Peaks and rapid variations of the flow variables will be smeared out and very fine grids are required to obtain accurate solutions [46]. Another straightforward approximation for ϕ_f can be derived under the assumption of a linear variation of ϕ between the nearest cell centers P and N, as shown in Fig. 4.2:

$$\phi_f = (1 - \lambda_x)\phi_N + \lambda_x \phi_P, \tag{4.12}$$

where λ_x is the interpolation factor defined as the ratio of distances $\overline{fN}$ and $\overline{PN}$:

$$\lambda_x = \frac{\overline{fN}}{\overline{PN}}. \tag{4.13}$$

The differencing or interpolation scheme using Eq. (4.12) is called *central differencing* (CD) and is of second-order accuracy even on non-uniform grids [46]. The major disadvantage of the CD is the generation of unphysical oscillations of the numerical solution for convection-dominated problems, thus violating the boundedness of the solution [64, 59, 46].

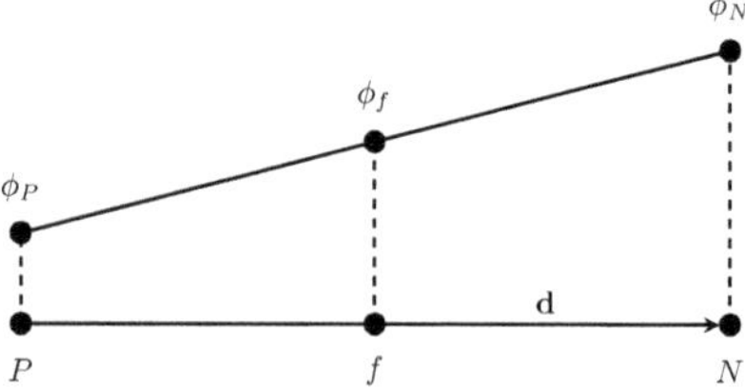

Fig. 4.2: Face interpolation of ϕ using the *central differencing* (CD) scheme.

In order to combine the advantages of the both schemes - namely the boundedness of the UD and the accuracy of the CD - many other attempts (e.g., the total variation dimishing

(TVD) and normalized variable diagram (NVD) schemes) have been made to find the most suitable face interpolation approach. The most high-order differencing schemes represent a linear combination or blending (*blending differencing*) of UD and CD:

$$\phi_f = (1-\gamma)\left(\phi_f\right)^{UD} + \gamma\left(\phi_f\right)^{CD}, \tag{4.14}$$

where γ is a blending factor depending on local flow values and/or the chosen flux limiter function (e.g., Sweby, Van Leer or QUICK). However, since the accuracy of the discretization is of highest priority for the reasonable application of large-eddy simulations, the CD scheme (4.12) was exclusively used in this work.

Diffusion term

The discretiztion of the diffusion term can be accomplished in a similar fashion by applying Eqs. (4.7) and (4.8):

$$\int_{V_P} \boldsymbol{\nabla} \cdot \left(\rho\Gamma_\phi \boldsymbol{\nabla}\phi\right)\mathrm{d}V = \int_S \mathrm{d}\mathbf{S} \cdot \left(\rho\Gamma_\phi \boldsymbol{\nabla}\phi\right) \approx \sum_f \left(\rho\Gamma_\phi\right)_f \mathbf{S} \cdot (\boldsymbol{\nabla}\phi)_f, \tag{4.15}$$

where $\left(\rho\Gamma_\phi\right)_f$ is assumed to be known. In case of an orthogonal mesh, i.e., vectors $\mathbf{d}$ and $\mathbf{S}$ are parallel (see Fig. 4.1), the following expression can be used for the evaluation of the surface-normal gradient:

$$\mathbf{S} \cdot (\boldsymbol{\nabla}\phi)_f = |\mathbf{S}|\frac{\phi_N - \phi_P}{|\mathbf{d}|}, \tag{4.16}$$

which is only of second-order accuracy for fully orthogonal grids. An alternativ approximation for non-orthogonal meshes consists of the calculation of the cell-center gradient for two cells sharing the corresponding face:

$$(\boldsymbol{\nabla}\phi)_P = \frac{1}{V_P}\sum_f \mathbf{S}\phi_f, \tag{4.17}$$

followed by a linear interpolation onto the cell face:

$$(\boldsymbol{\nabla}\phi)_f = f_x(\boldsymbol{\nabla}\phi)_P + (1-f_x)(\boldsymbol{\nabla}\phi)_N. \tag{4.18}$$

The discretization of the diffusion term using Eqs. (4.17) and (4.18) is of second-order accuracy, even for non-orthogonal meshes. However, similar to the central differencing (CD) for the discretized convective term, the boundedness of the numerical solution cannot be ensured and may lead to oscillatory solutions. Additionally, the truncation error is four times larger compared to Eq. (4.16) [64]. In order to prevent the higher accuracy and lower truncation error of Eq. (4.16) on non-orthogonal meshes, the product $\mathbf{S} \cdot (\boldsymbol{\nabla}\phi)_f$ can be splitted into two parts:

$$\mathbf{S}\cdot(\nabla\phi)_f = \underbrace{\boldsymbol{\Delta}\cdot(\nabla\phi)_f}_{\text{orthogonal contribution}} + \underbrace{\mathbf{k}\cdot(\nabla\phi)_f}_{\text{non-orthogonal contribution}}, \tag{4.19}$$

where the vectors $\boldsymbol{\Delta}$ and $\mathbf{k}$ have to fulfill the following condition:

$$\mathbf{S} = \boldsymbol{\Delta} + \mathbf{k}. \tag{4.20}$$

Vector $\boldsymbol{\Delta}$ is chosen to be parallel with $\mathbf{d}$. Thus, the orthogonal contribution can be evaluated using Eq. (4.16) while the less accurate approach is limited to the non-orthogonal part. As illustrated in Fig. 4.3, vector $\mathbf{d}$ can be found using the so-called *over-relaxed approach* [64], which is used in the present work:

$$\boldsymbol{\Delta} = \frac{\mathbf{d}}{\mathbf{d}\cdot\mathbf{S}}|\mathbf{S}|^2. \tag{4.21}$$

Finally, the discretized diffusion term can be written as:

$$\mathbf{S}\cdot(\nabla\phi)_f = |\boldsymbol{\Delta}|\frac{\phi_N - \phi_P}{|\mathbf{d}|} + \mathbf{k}\cdot(\nabla\phi)_f, \tag{4.22}$$

where $\boldsymbol{\Delta}$ can be evaluated from Eq. (4.21) and the cell face interpolation of $\nabla\phi$ is achieved using Eq. (4.18). It has to be mentioned that the application of non-orthogonal can lower or extinguish the boundedness of the solution. If boundedness is highly preferred, any non-orthogonal correction should be avoided, even though this step will reduces the accuracy.

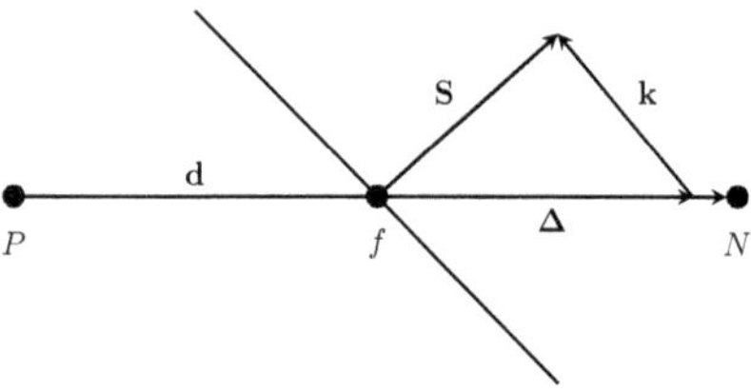

Fig. 4.3: Correction for non-orthogonal grids: *over-relaxed approach.*

Source term

Source terms, such as $S_\phi(\phi)$ of Eq. (4.3), can be a general function of ϕ. All terms of the generic transport equation that cannot be written as convection, diffusion or temporal terms are consequently treated as sources. Before the discretization, the source term needs to be linearized:

$$S_\phi(\phi) = S_E + S_I\phi, \tag{4.23}$$

where S_E and S_I may be functions of ϕ. The term is then integrated over a CV as follows:

$$\int_{V_P} S_\phi(\phi)\mathrm{d}V = S_E V_P + S_I V_P \phi_P. \tag{4.24}$$

The term S_E is treated in an explicit manner and will therefore appear on the right-hand side of the resulting linear algebraic system. The decision whether S_I is treated implicitly or explicitly depends on the its sign. For $S_I < 0$, this term will be handled implicitly to improve the diagonal dominance of the matrix, which improves the solver convergence [64]. In the case of $S_I > 0$, this term will be treated explicitly.

4.2 Temporal discretization

This sections briefly presents the applied spatial discretization procedure. As described in the previous section, the discretization of the spatial terms is accomplished in two major steps - the transformation of the surface and volume integrals into discrete sums and expressions that give the cell face values of the variable as a function of cell values. Considering Eqs. (4.10), (4.15), and (4.24), and assuming that the control volumes do not change in time, the spatially discretized form[4] of the general transport equation (4.5) can be rewritten as:

$$\int_t^{t+\Delta t} \left[\left(\frac{\partial \rho \phi}{\partial t} \right)_P V_P + \sum_f F \phi_f - \sum_f \left(\rho \Gamma_\phi \right)_f \mathbf{S} \cdot (\boldsymbol{\nabla} \phi)_f \right] \mathrm{d}t \\ = \int_t^{t+\Delta t} (S_E V_P + S_I V_P \phi_P) \mathrm{d}t. \tag{4.25}$$

In this work, the backward differencing in time is used for all conducted simulations. This temporal scheme is second-order accurate in time and neglects the temporal variation of the face values [64]. Accordingly, the approximation of the temporal derivative can be derived using three consecutive time levels n:

$$\frac{\partial \phi}{\partial t} = \frac{3\phi^n - 4\phi^{n-1} + \phi^{n-2}}{2\Delta t}. \tag{4.26}$$

The final form of the discretized transport equation (4.25) with backward differencing in time reads:

$$\frac{3\rho_P \phi^n - 4\rho_P \phi^{n-1} + \rho_P \phi^{n-2}}{2\Delta t} V_P + \sum_f F \phi_f^n - \sum_f \left(\rho \Gamma_\phi \right)_f \mathbf{S} \cdot (\boldsymbol{\nabla} \phi)_f^n \\ = S_E V_P + S_I V_P \phi_P^n. \tag{4.27}$$

[4]This expression is usually referred to as "semi-discretized" form of the general transport equation [59].

This produces a system of algebraic equations that must be solved for ϕ_P^n. Despite the fact that the boundedness of the solution cannot be guaranteed [64, 46], backward-differencing in time is preferred in the present work because of its simplicity and numerical stability as well as for its second-order accuracy in time. The latter one is a major requirement for the reliable application of large-eddy simulations. Moreover, the backward differencing in time offers the possibility of implicit or explicit temporal discretization. The corresponding size of time step Δt is generally restricted by the Courant number, or more precisely to the Courant-Friedrichs-Lewy (CFL) condition:

$$Co = \frac{\mathbf{u}(\mathbf{x}, t)\Delta t}{\Delta \mathbf{x}} \leq Co_{max}, \tag{4.28}$$

which represents a fundamental relation of the time step Δt to the characteristic convection time $\Delta \mathbf{x}/\mathbf{u}(\mathbf{x}, t)$, the time required for a disturbance to be convected over a distance $\Delta \mathbf{x}$ [46]. Hence, the CFL condition prescribes the (maximum) distance which can be overcome by a certain quantity in a given time step. The stable application of explicit temporal schemes usually requires $Co_{max} \leq 1$ [151, 46], whereas implicit approaches are less sensitive concerning numerical instabilities, resulting in higher maximum Courant numbers or larger time steps respectively. Although backward differencing in time is (theoretically) not restricted to the condition $Co_{max} \leq 1$, the Courant number in the presented simulations was adjusted to $0.3 \leq Co \leq 0.5$, to ensure a high stability and a sufficient accuracy in time.

4.3 Pressure-velocity coupling

The pressure-velocity coupling[5] applied in this work is the so-called PISO (Pressure-Implicit with Splitting of Operators) algorithm proposed by Issa [63] and will be discussed at this point.

For the derivation of the pressure equation and the calculation of a divergence-free velocity field, a semi-discretized form of the momentum equation is used as starting point:

$$\mathcal{A}_D \mathbf{u} = \mathcal{A}_H - \boldsymbol{\nabla} p. \tag{4.29}$$

Following the procedure proposed by Rhie and Chow [122], the pressure gradient term is not discretized at the current stage. This equation is obtained from the integral from of the momentum equation (2.6) and the application of the discretization methods described previously. The $\mathcal{A}_H$ term inculdes the matrix coefficients for all neighbors multiplied by the corresponding velocities as well as all remaining source terms excluding the pressure gradient. Rearrangement of Eq. (4.29) leads to the following expression for the velocity

[5]For incompressible flows, the pressure cannot be directly evaluated through the coupled mass and momentum equation, Eqs. (2.4) and (2.6), since the mass conservation represents only a kinematic condition for the velocity field. To overcome this issue, an additional pressure equation is derived which is then solved by an appropriate pressure-velocity coupling approach (e.g., SIMPLE, PISO or PIMPLE).

field:

$$\mathbf{u} = \frac{\mathcal{A}_H}{\mathcal{A}_D} - \frac{1}{\mathcal{A}_D}\boldsymbol{\nabla} p, \tag{4.30}$$

where the velocities on the cell faces f can be expressed by face interpolation of Eq. (4.30) as:

$$\mathbf{u}_f = \left(\frac{\mathcal{A}_H}{\mathcal{A}_D}\right)_f - \left(\frac{1}{\mathcal{A}_D}\right)_f (\boldsymbol{\nabla} p)_f. \tag{4.31}$$

which will be used later to calculate the face fluxes F. Using the discretized form of the continuity equation (2.4):

$$\boldsymbol{\nabla} \cdot \mathbf{u} = \sum_f \mathbf{S} \cdot \mathbf{u}_f = 0, \tag{4.32}$$

and substitution of Eq. (4.31) into Eq. (4.32) leads to:

$$\boldsymbol{\nabla} \cdot \left(\frac{1}{\mathcal{A}_D}\boldsymbol{\nabla} p\right) = \boldsymbol{\nabla} \cdot \left(\frac{\mathcal{A}_H}{\mathcal{A}_D}\right) = \sum_f \mathbf{S} \cdot \left(\frac{\mathcal{A}_H}{\mathcal{A}_D}\right)_f, \tag{4.33}$$

where the Laplacian term on the left-hand side of Eq. (4.33) is discretized as described in section 4.1. The final form of the discretized, incompressible Navier-Stokes system is:

$$\mathcal{A}_D \mathbf{u} = \mathcal{A}_H - \sum_f \mathbf{S}(p)_f, \tag{4.34}$$

$$\sum_f \mathbf{S} \cdot \left[\left(\frac{1}{\mathcal{A}_D}\right)_f (\boldsymbol{\nabla} p)_f\right] = \sum_f \mathbf{S} \cdot \left(\frac{\mathcal{A}_H}{\mathcal{A}_D}\right)_f. \tag{4.35}$$

The face fluxes F, which satisfies Eq. (4.32) by construction are evaluated as:

$$F = \mathbf{S} \cdot \mathbf{u}_f = \mathbf{S} \cdot \left[\left(\frac{\mathcal{A}_H}{\mathcal{A}_D}\right)_f - \left(\frac{1}{\mathcal{A}_D}\right)_f (\boldsymbol{\nabla} p)_f\right]. \tag{4.36}$$

Based on this calculus, the PISO algorithm solves the momentum equation, Eq. (4.34), in a first step (*momentum predictor*) where the pressure field from the previous time step is used. It follows an approximation of the new velocity field, which is used to assemble the $\mathcal{A}_H$ operator and to formulate and solve the pressure equation (*pressure solution*). From Eq. (4.36), a set of conservative fluxes, consistent with the new pressure field, is calculated. As a consequence of the new pressure field, the velocity field is corrected using Eq. (4.30) (*explicit velocity correction*).

4.4 Boundary conditions

The numerical solution of the governing equations of fluid motions and heat transfer, Eqs. (2.4), (2.6), and (2.14), requires the definition of appropriate initial and boundary conditions. For transient flows, the initial conditions are usually generated by analytical solutions, mapping techniques (i.e., precursor calculations, recycling methods) or inflow generators (synthetic turbulence). However, from a pure mathematical point of view, most of the existing boundary conditions can be categorized as follows[6]:

- **Dirichlet (or fixed value) boundary condition** prescribes the value of $\phi(\mathbf{x}, t)$ at the boundary face b to be $\phi_b(\mathbf{x}, t)$:

$$\phi(\mathbf{x}, t) = \phi_b(\mathbf{x}, t). \tag{4.37}$$

- **Neumann (or fixed gradient) boundary condition** prescribes the value of gradient $\nabla\phi(\mathbf{x}, t)$ normal to the boundary face b to be $f_b(\mathbf{x}, t)$:

$$\left.\frac{\partial\phi(\mathbf{x}, t)}{\partial\mathbf{n}}\right|_b = f_b(\mathbf{x}, t) \quad \text{or} \quad \left(\frac{\mathbf{S}}{|\mathbf{S}|}\cdot\nabla\phi(\mathbf{x}, t)\right)_b = f_b(\mathbf{x}, t). \tag{4.38}$$

The choice of the boundary conditions strongly depends on the flow variables and the physical meaning of the associated boundary face. In the following, a brief overview of all applied boundary conditions is presented, whereas further specifications of the boundary conditions are described later in the corresponding sections of each flow configuration.

Inlet boundary

The flow velocity field and all scalar variables (e.g., turbulence kinetic energy, temperature) are prescribed at the inlet boundary. For consistency, the pressure boundary condition is zero gradient.

Outlet boundary

In order to satisfy the overall mass balance for the computational domain by the pressure equation (see section 4.3), the fixed value boundary condition is applied for pressure with a zero gradient boundary condition on the velocity. For strong backflow areas, the stability of the simulation can be further increased by special outlet boundary conditions which prevents an inflow through a boundary specified as outlet.

[6]All remaining types of boundary conditions (i.e., Robin, Cauchy and mixed boundary conditions) are simply combinations of the Dirichlet and Neumann boundary conditions.

Impermeable no-slip walls

The flow velocity field at the wall is fully prescribed by the no-slip condition, i.e., all velocity components as well as the turbulence kinetic energy are set to be zero. The pressure boundary condition is zero gradient. In case of heating and/or cooling, the temperature is additionally given at the wall.

Periodic boundary

Flows with homogeneous characteristics in a certain direction can be treated as periodic. Therefore, despite their spatial distance, the boundary faces on the opposite periodic boundary are numerically treated as neighboring (internal) cell faces.

Mapping plane boundary

The mapping plane boundary, often referred to as recycling method, is an extension of the inlet boundary where the simulation generates its own turbulent inflow conditions. Thus, the instantaneous flow variables are recorded in an interior plane located downstream[7] of the inflow. These data are then rescaled and reintroduced at the inlet boundary [91, 124].

4.5 Numerical procedure of the applied Lagrangian particle tracking

The equations of particle motion, Eqs. (3.14) and (3.15), are numerically solved using a Lagrangian particle tracking (LPT) approach based on the *TrackToFace*[8] algorithm proposed by Macpherson et al. [93]. This method is proven to be robust, computationally efficient and can be employed for particle tracking even in complex three-dimensional domains and for unstructured polyhedral meshes (including mesh deformation and motion). It is suitable for a wide range of engineering problems ranging from injected fuel sprays in internal combustion engines [112] to molecular dynamics of nano scale liquid systems [94].

Following the idea of the TrackToFace algorithm, the forces acting on the particle are determined through the averaged, cell-based values of the local fluid properties which have to be interpolated at the specific particle position. Based on these forces, the final position of the particle at the end of the time step is evaluated and a virtual displacement vector is computed. The particle is then moved only for a fraction of the time step long enough to track the particle to the next cell face it has to cross[9]. At the cell face position, the fluid

[7] An appropriate distance between inlet boundary and mapping plane depends significantly on the flow type. However, it must be chosen in a way that the decay of the autocorrelation function of the velocity fluctuation is guaranteed [116].

[8] *TrackToFace* is the name of the corresponding OpenFOAM function. Macpherson et al. [93] originally introduced this algorithm without any specific name.

[9] If the particle does not cross a new cell face within the time step, the particle is still inside the same cell and can be moved directly to the end point of the motion.

properties of the new grid cell is interpolated to the particle position and the new particle forces as well as the virtual displacement vector for the remaining time step duration is calculated. The particle tracking is stopped at the end of the time step, whereby the particle is located inside a cell and not on a cell face. Since this procedure allows a straightforward evaluation of the residence time of the particle in each computational cell, the momentum exchange or back force for the two-way coupling can be easily determined (see Eq. 3.19). Moreover, the TrackToFace algorithm allows the tracking of parcels (see section 3.2), representing a group of physical particles with identical properties (e.g., position, diameter and velocity) whereas an interaction of the particles inside a parcel is neglected. The concept of parcels is important for flow configurations with a high mass loading, since the number of particles in this case usually exceeds the computational tracking capabilities.

5 Modeling of particulate fouling on structured heat transfer surfaces

This chapter summarizes the development of a noval Eulerian-Lagrangian approach, suitable for the spatial- and time-resolved simulation of particulate fouling on structured heat transfer surfaces in turbulent flows. Most of the work presented here was originally published in Kasper et al. [67] and Deponte et al. [34].

5.1 Eulerian-Lagrangian approach

Particulate fouling on heat transfer surfaces is a complex multiphysical problem. It includes different steps (e.g., mass transport, attachment and formation, removal and resuspension) as described at the beginning in section 1.2, which have to be considered in the numerical simulation to obtain reliable results. Therefore, a multiphase Eulerian-Lagrangian approach was developed, mainly consisting of the following two branches:

1. **Eulerian branch:** The flow fields (i.e., $\mathbf{u}$, p and T) of the turbulent carrier flow (continuous phase) are obtained from large-scale resolving LES:

$$\nabla \cdot \overline{\mathbf{u}} = 0, \tag{5.1}$$

$$\frac{\partial \overline{\mathbf{u}}}{\partial t} + \nabla \cdot (\overline{\mathbf{u}}\,\overline{\mathbf{u}}) = -\nabla \overline{p} + \nu \nabla^2 \overline{\mathbf{u}} - \nabla \cdot \boldsymbol{\tau}_{SGS} - \overline{\mathbf{S}}_p\big(\alpha_f\big) + \overline{\mathbf{S}}_M, \tag{5.2}$$

$$\frac{\partial \overline{T}}{\partial t} + \nabla \cdot \big(\overline{\mathbf{u}}\overline{T}\big) = \nabla \cdot \big(a\big(\alpha_f\big)\nabla \overline{T}\big) - \nabla \cdot \mathbf{q}_{SGS}, \tag{5.3}$$

   where $\overline{\mathbf{S}}_M$ is the additional source term contribution due to two-way momentum coupling, Eq. (3.19), and $\overline{\mathbf{S}}_p\big(\alpha_f\big)$ is a porosity source term contribution, based on the so-called *fouling phase fraction* α_f, which is further explained in section 5.1.1. It needs to be highlighted that the energy equations, Eq. (5.3), is expressed in terms of the fouling phase fraction (i.e., $a\big(\alpha_f\big)$) to account for the additional thermal resistance due to fouling deposits.

2. **Lagrangian branch:** Mass transport of the foulant particulates (dispersed phase) to the heat transfer surfaces (Eq. (3.14 and (3.15)), based on the spatial- and time-resolved carrier flow fields. Formation of fouling deposits based on an energy balance (derived from DLVO theory) as well as removal and resuspension of deposits due to local shear stresses.

In addition, a phase conversion algorithm is introduced which converts deposited particles into an additional porous continuous phase (fouling layer). This allows a significantly reduction of the computational effort since converted particles are removed from the costly particle-tracking algorithm, i.e., the amount of tracked particles is kept constant during the simulation. Furthermore, a multiscale modeling approach based on the heterogeneous multiscale method (HMM) [38, 37] was derived, which allows the simulation and assessment of technical relevant long-term fouling intervals (e.g., hours, weeks).

5.1.1 Formation of fouling deposits

The physical modeling of the foulant particle deposition on heat transfer surfaces is probably the most important and crucial part in the presented fouling simulations. Due to the fact that particle deposition is mainly caused by particle-wall adhesion within this work (this assumption is applicable to particle diameters of $D_p \leq 100\,\mu$m [45]), the implemented deposition model is based on the work Heinl and Bohnet [56] who investigated the conveying of quartz particles in horizontal pipes. Following the DLVO theory (named after Derjaguin, Landau, Verwey and Overbeek [36, 152]), this model is based on a energy balance around the particle-wall and particle-fouling layer collision and the deformation of the contact partner. According to Wang et al. [158], who studied the transport and adhesion of glass beads with different sizes in channels, the influence of electrostatic forces is only of importance for larger particles ($D_p \geq 50\,\mu$m) and are negligible small when the channel wall is already covered with particulates. Since only fine foulant particles (i.e., $D_p < 50\,\mu$m) were considered in this work, the local energy balance reads:

$$E_{kin,1} = E_{vdW} + E_{kin,2} + E_l, \tag{5.4}$$

where E_{kin} is the kinetic energy before (1) and after (2) wall collision, respectively, and E_{vdW} is the energy ratio which accounts for adhesion due to the van der Waals forces. The energy loss of a particle due to collision (i.e., plastic deformation and friction) is described by E_l and is eliminated by introducing the (plastic) coefficient of restitution e as:

$$e^2 = \frac{E_{kin,1} - E_l}{E_{kin,1}}. \tag{5.5}$$

The kinetic energy of a particle is a function of the particle diameter D_p, mass density ρ_p and the magnitude of the particle velocity u_p:

$$E_{kin} = \rho_p \frac{\pi}{12} D_p^3 u_p^2. \tag{5.6}$$

The van der Wals energy that is stored in the deformation of the particle, consists of an elastic part $E_{vdW,el}$ and a plastic part $E_{vdW,pl}$. Since the energy conversion due to elastic deformation is smaller than 1% [21], the total dissipation energy can be approximated as:

$$E_{vdW} = E_{vdW,el} + E_{vdW,pl} \approx E_{vdW,pl}. \tag{5.7}$$

The van der Waals pressure is given as [153, 154]:

$$p_{vdW} = \frac{\hbar\varpi}{8\pi z^3} G, \tag{5.8}$$

where G is the geometry factor, which is the diameter of the spherical particle, and $\hbar\varpi$ is the Lifshitz-van der Waals constant. The van der Waals energy E_{vdW} is found by integrating the product of the van der Waals pressure p_{vdW} and the deformed area ($A_{def} = \pi r_0^2$) over the distance to the wall z [60]:

$$E_{vdW} = \int_{z_0}^{\infty} p_{vdW} \cdot \pi r_0^2 \cdot \mathrm{d}z\,, \tag{5.9}$$

where r_0 is the radius of the deformed area. For small deformations ($h \leq D_p$), the contact area πr^2 is approximated in terms of the depth of the penetration h [21]:

$$\pi r^2 = \pi D_p \cdot h. \tag{5.10}$$

The energy needed for the deformation can be calculated by integrating the product of the yield stress (or strength) of the contact partner p_{pl} and area between 0 and the plastic deformed height h_{pl} (i.e., the depth of the penetration h is approximated by the depth of the plastic deformation h_{pl}) [60]:

$$E_{vdW,pl} = \int_{0}^{h_{pl}} p_{pl}\pi D_p \cdot h\,\mathrm{d}h = \frac{1}{2} p_{pl}\pi D_P h_{pl}^2. \tag{5.11}$$

For purely plastic deformation and considering only the van der Waals force, the depth of the penetration is found from the equilibrium between the plastic part of the van der Waals force, Eq. (5.11), and the kinetic energy, Eq. (5.6), excluding the energy loss due to elastic deformation:

$$h_{pl} = u_p D_P \sqrt{\frac{\rho_P}{6 p_{pl} e^2}}, \tag{5.12}$$

in which e is the plastic coefficient of restitution, Eq. (5.5). Substitution of Eq. (5.9) into (5.12) yields the total van der Waals energy:

$$E_{vdW} = \frac{\hbar\varpi}{16\pi z_0^2} D_p^2 u_P \sqrt{\frac{\rho_P}{6 p_{pl} e^2}}. \tag{5.13}$$

Recalling that the prerequisite of adhesion is that the particle is not able to leave the wall after wall collision (i.e., $E_{kin,2} = 0\,\mathrm{J}$), a critical particle velocity or sticking velocity can be derived through Eqs. (5.4) and (5.13):

$$u_{p,crit} = \sqrt{\left(\frac{\hbar\varpi}{e D_p 4\pi^2 z_0^2}\right)^2 \frac{3}{4 p_{pl}\rho_p}}. \tag{5.14}$$

where $\hbar\varpi$ is the Lifshitz-van der Waals energy, z_0 is the distance at contact, p_{pl} is the yield stress (strength) of the contact wall and e is the coefficient of restitution. The condition of particle sticking is achieved, if the particle velocity before the wall collision (impact velocity) is smaller than the critical particle velocity:

$$|\mathbf{u}_p| \leq u_{p,crit}. \tag{5.15}$$

To increase the computational efficiency of the presented approach, particles which fulfill the adhesion condition, Eq. (5.15), are converted into an additional solid phase (i.e., fouling deposits) and will be deactivated within the particle tracking algorithm. Thus, the amount of particles is kept nearly constant during the simulation which reduces the computational time enormously. The initiated fouling volume or fouling phase fraction α_f (based on Eq. (3.1))

$$\alpha_f = \frac{\sum_i V_{p,i}}{V_c} \tag{5.16}$$

relates the summarized volume $\sum_i V_{p,i}$ of the deposited monodisperse particles inside the considered computational grid cell to the volume of the mesh cell V_c itself and varies from 0 (no fouling within the cell) to 1 (the cell is completely filled by particulate fouling). Fig. 5.1 shows the basic concept of the implemented phase conversion algorithm.

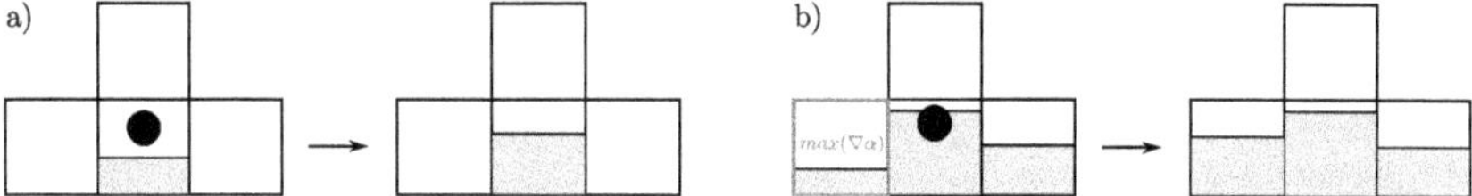

Fig. 5.1: Basic principle of the phase conversion algorithm: a) $V_c > V_p$ and b) $V_c < V_p$

There are two different cases for a deposited particle being converted into the fouling phase, as shown in Fig. (5.1). If the remaining local cell volume is greater than the particle volume, the new phase fraction α can be simply determined using Eq. (5.16). If the remaining local cell volume is smaller than the particle volume, the phase fraction is allocated to the neighbor cell with the maximum cell-based phase fraction gradient $\max(\nabla\alpha)$. Hence, the neighbor cell with the lowest phase fraction is filled with the fouling phase. To consider the influence of the fouling phase, an additional porosity source term (based on Darcy's law)

$$\overline{\mathbf{S}}_p(\alpha_f) = \alpha_f \frac{\mu_f}{K} \overline{\mathbf{u}}_f, \tag{5.17}$$

has been introduced into the momentum balance equation, Eq. (5.2), where K is the isotropic permeability of the fouling phase. Thus, the blocking effect or flow section contraction due to deposited particles is not explicitly considered within the calculations, but rather is modeled implicitly in terms of a porous fouling layer, as shown in Fig. 5.2.

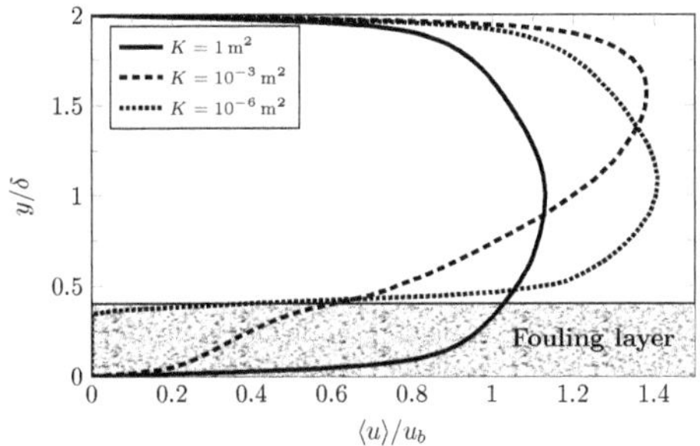

Fig. 5.2: Influence of the isotropic permeability K of the fouling layer on the time-averaged streamwise velocity $\langle u\rangle/u_b$ in a turbulent channel flow ($Re_\tau = 395$).

Furthermore, any physical property φ_i (e.g., density, dynamic/kinematic viscosity or thermal diffusivity) for partially filled cells is interpolated as follows:

$$\phi_c = \alpha\phi_{fouling} + \left(1 - \alpha_f\right)\phi_{fluid}, \tag{5.18}$$

whereas the physical properties of the carrier fluid are fully applied at cells without fouling phase ($\alpha_f = 0$) and cells which are completely occupied by the fouling phase ($\alpha_f = 1$) take the physical properties of the fouling material. This procedure allows the evaluation of the heat transfer under consideration of particulate fouling (e.g., fouling resistance R_f) and prevents the solving of an transport equation for the fouling phase as well as the application of costly remeshing procedures. Aging of fouling deposits is not considered in the present model, i.e. the strength of the fouling layer does not change with time.

5.1.2 Removal and resuspension of fouling deposits

Removal and resuspension (or reentrainment) of fouling material occurs immediately after the first formation of deposits. It is governed by particle-surface interaction (i.e., contact forces) and particle-fluid interactions (i.e., aerodynamic/hydrodynamic forces). The equilibrium between these forces determines whether a particle adhere to the surface or are resuspended [57]. Numerous models were theoretically derived in order to capture and explain resuspension of particles in turbulent flows, which can be roughly categorized in *models based on force balance* and *models based on energy accumulation* [173]. In the simplest force balance model, introduced by Wen and Kasper [164], a particle is resuspended if the instantaneous lift force is larger than the surface adhesive model. The RRH model, proposed by Reeks, Reed and Hall [120], is an energy accumulation based model in which a particle is resuspended when it has accumulated enough vibrational energy from the turbulent flow to detach itself from the surface.
In this work, it is assumed that the removal and reentrainment process of fouling deposits is driven by local shear stresses. Therefore, a simple linear removal model was derived

based on the suggestions of Kern and Seaton [69] and Taborek et al. [142]:

$$\alpha_r = \min\left(\frac{|\tau_c|}{\tau_{rel}}\frac{V_p}{V_c}, \frac{V_p}{V_c}\right), \tag{5.19}$$

where τ_{rel} is a relative shear stress and τ_c is the local cell-based shear stress. The relative shear stress has to be derived experimentally or theoretically (see e.g., Bobe [14]) and can be interpreted as a strength threshold value at which the removal and reentrainment of fouling deposits in each grid cell starts. The number of resuspended spherical mono-dispersed particles can be determined using the definition of the sphere volume:

$$n = \frac{\alpha_r V_c}{\pi D_p^3/6}. \tag{5.20}$$

The resuspended fouling particles are released into the turbulent carrier flow at the corresponding grid cells and will be tracked through the computational domain based on the LPT approach described in chapter 3. Fig. 5.3 shows a flow chart, which summarizes the proposed multiphase Eulerian-Lagrangian approach for the numerical simulation of particulate fouling on structured heat transfer surfaces.

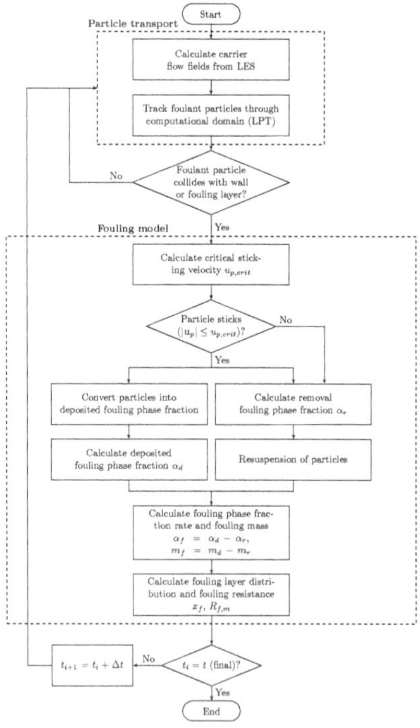

Fig. 5.3: Flow chart of the proposed multiphase Eulerian-Lagrangian approach.

5.2 Multiscale modeling for the simulation of long-term fouling intervals

The formation of fouling layers in industrial applications, which are in a state of equilibrium ($\frac{\mathrm{d}m_f}{\mathrm{d}t} = 0\,\frac{\mathrm{kg}}{\mathrm{m}^2\mathrm{s}}$), is usually a process occurring on large time scales (e.g., days, weeks, months, or even years). By contrast, the numerical time steps during the simulations are strongly restricted by the Courant-Friedrichs-Lewy (CFL) condition, Eq. (4.28), to ensure convergence and stability while solving the underlying partial differential equations numerically. As mentioned earlier in section 4.2, the stable application of explicit temporal schemes usually requires $Co_{max} \leq 1$, whereas implicit approaches are less sensitive concerning numerical instabilities, resulting in higher maximum Courant numbers or larger time steps, respectively. Although the implicit backward differencing is applied in this work, the Courant number is restricted to $0.3 \leq Co \leq 0.5$ in all simulations for higher stability and sufficient accuracy in time, resulting in time steps down to $\Delta t = 10^{-6}\,\mathrm{s}$ for the investigated fouling configurations. Hence, the time step of the fluid flow is many orders of magnitude smaller than the real fouling process, which is a classical multiscale problem. Simulation of the whole fouling time period is impossible due to the massive computational effort and missing applicability; therefore, an approach based on the heterogeneous multiscale method (HMM) [1, 38, 37] is introduced to overcome this issue by connecting both time scales and to evaluate long-term fouling intervals using Eulerian-Lagrangian LES.

5.2.1 Capturing of the macroscale and microscale fouling behavior

Following the basic idea of the HMM, the behavior of the fouling phase fraction α_f is roughly estimated by an (incomplete) macroscale model, where specific details of the model are missing. At this point, recall that α_f is directly related to the local (cell-based) fouling mass m_f and fouling layer thickness x_f. A microscale model, which describes the behavior of the fluid flow and the particle transport, as well as the particle deposition and resuspension, is used during the simulation to supply all the data needed to complete the macroscale solver. Fig. 5.4 presents the coupling between both models.

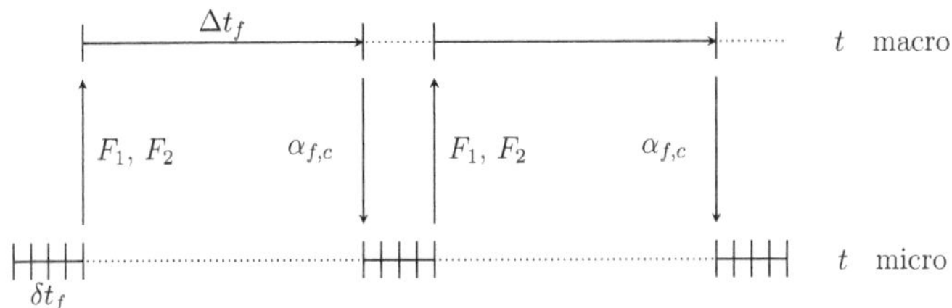

Fig. 5.4: Scheme of the coupling between macroscale and mircoscale model.

It is realized in such way that the macro state provides the constraints for setting up the micro model, while the micro model provides the needed input data for the macro model [37]. The macro model requires the reinitialization of the microscale solver at each macro time step. The occurrence of particulate fouling in tubes and channels usually causes a flow section contraction that results in higher flow velocities, an enhancement of the wall shear stresses, and fouling removal. After a certain time, the solid removal per unit time and surface area is equal to the deposited solids, where the fouling mass and fouling resistance reach limiting values [15]. At the macroscopic level, the characteristic fouling layer evolution can be assumed to be nearly estimated for each partially filled computational grid cell in terms of the fouling phase fraction α_f by a simple growth model, based on different theoretical approaches [15, 45, 17] describing the asymptotic behavior of fouling deposits for closed systems with restricted cross-sections (e.g., pipes or channels):

$$\alpha_{f,c} = F_1 + F_2 \ln(t), \tag{5.21}$$

where F_1 and F_2 are missing input parameters and have to be supplied by the microscale model. These input data are estimated by a regression analysis based on the least squares method. The corresponding microscale solver is the multiphase Eulerian-Lagrangian approach, as shown in Fig. 5.3. Thus, the proposed multiscale modeling procedure is performed in the following steps (see also Fig. 5.4 and 5.5):

1. Microscale simulation of the micro fouling time interval $\delta t_{f,i}$, based on the Eulerian-Lagrangian approach.

2. Initial extraction of the macroscale solver input parameters $F_{1,i}$ and $F_{2,i}$ for each computational grid cell using regression analysis (least squares method).

3. Prediction of the cell-based fouling phase fraction $\alpha_{f,c}$ for the macro fouling time interval $\Delta t_{f,i}$ using Eq. (5.21) and the initial input parameters $F_{1,i}$ and $F_{2,i}$. If $\alpha_{f,c} > 1$, the remaining fouling phase fraction is allocated to the neighboring empty or partially filled grid cells.

4. Reinitialization of the microscale solver flow fields (e.g. velocity, temperature, turbulence kinetic energy) and simulation of the next micro fouling interval $\delta t_{f,i+1}$. The microscale simulation is now constrained, so the predicted fouling phase $\alpha_{f,c}$ from the macroscale solver is equal to the initial fouling phase $\alpha_{f,c}$ at the beginning of the current microscale simulation.

5. Update of the macroscale solver input parameters to $F_{1,i+1}$ and $F_{2,i+1}$ for each computational grid cell based on the current and previous microscale simulation(s).

6. Prediction of the cell-based fouling phase fraction $\alpha_{f,c}$ for the next fouling time interval $\Delta t_{f,i+1}$ using Eq. (5.21) and the updated parameter $F_{1,i+1}$ and $F_{2,i+1}$. If $\alpha_{f,c} > 1$, the remaining fouling phase fraction is allocated to the neighboring empty or partially filled grid cells.

7. Repetition of step 4-6 until quasi-static fouling conditions are obtained $\left(\frac{\mathrm{d}m_f}{\mathrm{d}t} \approx 0\,\frac{\mathrm{kg}}{\mathrm{m^2 s}}\right)$.

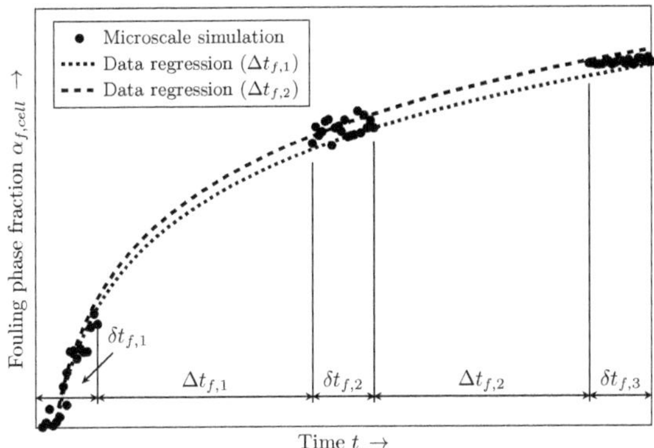

Fig. 5.5: Exemplary prediction of the cell-based fouling phase fraction $\alpha_{f,c}$ based on the proposed multiscale approach.

A shortcoming of the presented multiscale modeling approach is that it is based on a preconceived macroscale model. Therefore, special attention has to be paid to the choice of a suitable macroscale model that will produce accurate numerical fouling predictions. Another critical point is the selection of the micro fouling time interval δt_f, as this significantly affects the overall accuracy of the multiscale modeling. In this work, the microscale fouling time interval δt_f is chosen under the condition that at least one complete cycle of the asymmetric vortex switching, which is the main driving force for the removal and resuspension of particulate deposits [66], is realized within the microscale simulation. Further explanations regarding the determination of the vortex switching frequency and the choice of the microscale time interval are presented in section 7.1.2.

5.2.2 Application of the multiscale approach

Fig. 5.6 also shows the progression of the mass-based fouling resistance $R_{f,m}$ (see Eq. (7.10)) around the single spherical dimple with a dimple depth-to-dimple diameter ratio of $t/D = 0.26$, as predicted by a pure microscale simulation and the presented multiscale approach. Both fouling curves are close to each other and reveal the characteristic asymptotic growth of the fouling resistance against a comparable asymptotic value $R^*_{f,m}$. In the multiscale simulation, only 1.5 min of the total physical fouling time is calculated using the extremely costly Eulerian-Lagrangian LES. This leads to enormous savings in computational time when compared with the classical approach (Fig. 5.3), but with almost identical results. Although the influence of the initial conditions and the choice of the micro/macro fouling time intervals on the accuracy of the multiscale approach needs further investigation, this approach provides a promising opportunity to evaluate long-term fouling evolution (e.g., for industrial applications) and to increase the applicability the proposed numerical fouling approach.

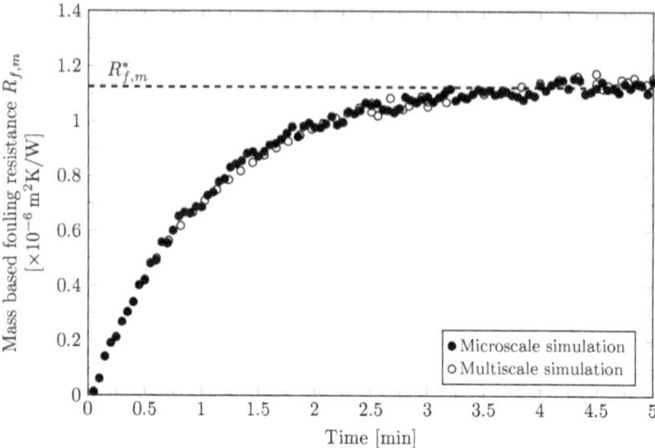

Fig. 5.6: Progression of the mass-based fouling resistance around the single dimple $t/D = 0.26$ for $u = 1\,\mathrm{m/s}$ and $C = 2\,\mathrm{g/L}$ obtained from microscale and multiscale simulations.

6 Validation

In this chapter, the validation of the implemented numerical methods is presented which ensures and highlights the proper functionality and high accuracy of the proposed Eulerian-Lagrangian approach. The validation of Lagrangian branch (i.e., point-particle tracking, particle deposition) and the Eulerian branch (i.e., LES of the carrier flow field) has been done for the following canonical test cases:

1. Particle-laden Taylor-Green vortex flow
2. Particle-laden backward-facing step flow
3. Particle-laden flow in a simplified combustion chamber
4. Particle-laden turbulent channel flow

The presented work, except the study of the dispersed two-phase flow in the simplified combustion chamber (section 6.3), was originally published in Kasper et al. [67, 68].

6.1 Particle-laden Taylor-Green vortex flow

The two-dimensional Taylor-Green vortex flow is chosen to assess the basic performance of the implemented LPT algorithm. This flow is selected due to the existence of its exact solution of the corresponding instantaneous velocity field and stream function as a special case of a two-dimensional time dependent solution of the Navier-Stokes equations.

6.1.1 Case description and numerical setup

The two-dimensional Taylor-Green vortex is assumed to be in the $x-y$ plane, and the flow is uniform in z direction. Thus, the instantaneous local stream function can be written as [165]:

$$\boldsymbol{\Psi}(x,\,y,\,t) = \frac{\omega_0}{\kappa^2}\cos\left(\kappa_x x\right)\cos\left(\kappa_y y\right)\exp\left(-Re_0^{-1}\kappa^2 t\right), \tag{6.1}$$

where the corresponding instantaneous fluid velocity components can be directly derived from Eq. (6.1):

$$u_x = \frac{\partial \boldsymbol{\Psi}}{\partial y} = -\omega_0\frac{\kappa_y}{\kappa^2}\cos\left(\kappa_x x\right)\sin\left(\kappa_y y\right)\exp\left(-Re_0^{-1}\kappa^2 t\right) \tag{6.2}$$

$$u_y = -\frac{\partial \Psi}{\partial x} = \omega_0 \frac{\kappa_x}{\kappa^2} \sin(\kappa_x x) \cos(\kappa_y y) \exp\left(-Re_0^{-1} \kappa^2 t\right) \tag{6.3}$$

where ω_0 is the initial vorticity maximum, κ_x and κ_y are the wave numbers in x- and y-directions and $\kappa^2 = \kappa_x^2 + \kappa_y^2$. It is assumed that the gravitational acceleration is normal to the flow plane $(x - y)$, so that it does not affect the particle dynamics. The simulations are performed with the following parameters: $Re_0^{-1} = 0.004$, $\kappa_x = \kappa_y = 1$, $\omega_0 = 2$, x and y are in the range of 0 and 2π m. Spherical particles with a constant diameter of $D_p = 0.001$ m were randomly seeded within the flow field.

6.1.2 Influence of the particle response time on particle dynamics

The primary objective of this simple test case is to examine effects of the particle or momentum response time τ_p, Eq. (3.10), on the particle trajectories considering Stokes drag ($Re_p < 1$), which exemplifies the basic functionality of the implemented LPT algorithm in a qualitative manner. In the case of nearly massless particles (i.e., $\tau_p \approx 0$ s), the particle trajectories should follow the flow streamlines exactly. For the case of heavier particles ($\tau_p > 0$ s), their trajectories are expected to deviate from the flow streamlines due to inertial effects. Only the drag force is considered within the simulations. Fig. 6.1 compares the particle trajectories after simulating 3.4 s of physical real time for two different momentum response times to the streamlines of the two-dimensional Taylor-Green vortex computed from stream function, Eq. (6.1).

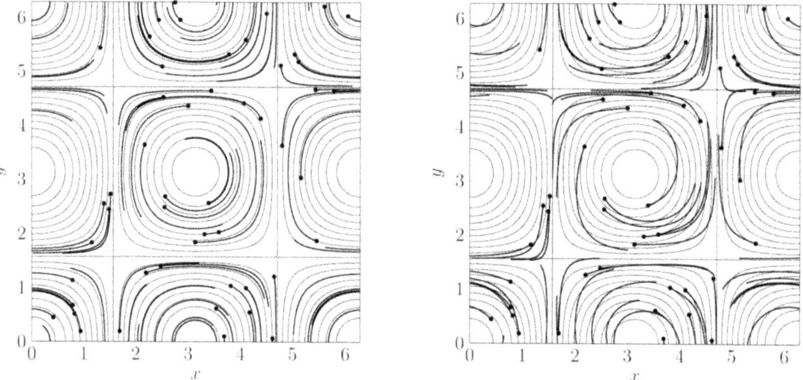

Fig. 6.1: Two-dimensional Taylor-Green vortex: streamlines (contour lines) and particle trajectories (thick dotted lines) for $\tau_p \approx 0$ s (left) and $\tau_p \approx 10$ s (right); thick dots represents the initial positions of the randomly seeded particles.

It is obvious that the motion of the particles seems to be correctly described by the LPT approach, since the nearly massless particles ($\tau_p \approx 0$ s) follow the streamlines very well. With increasing particle response time ($\tau_p \approx 10$ s), the particles get their own inertia and they lose the ability to follow the carrier flow. As expected, our numerical results show that an increase of the particle response times τ_p leads to an increased deviation of the particle trajectories from the flow streamlines. The obtained results confirm that the employed LPT routines are capable to describe the basic motion of the particles correctly.

6.2 Particle-laden turbulent backward-facing step flow

The particle-laden turbulent backward-facing step flow (BFS) is chosen for a more thorough, complex assessment of the implemented particle-tracking algorithm. Major aspects of this well-known canonical test case are the separation of the main flow, the development of a shear layer and reattachment of the boundary layer approximately at $x/H = 7$ behind the step, which makes this flow configuration an ideal benchmark for turbulence models and turbulent dispersed multiphase flows. Among several experimental and numerical studies of particle-laden backward-facing step flows, the laser Doppler anemometry (LDA) measurements of Fessler and Eaton [47] are used to evaluate the reliability of the LPT for fully turbulent flows using LES. Moreover, the general performance of three subgrid-scale models (DOE, DSM and SM) as well as their influence on the predicted particle motion is evaluated.

6.2.1 Case description and numerical setup

The computational domain is chosen according to the experimental setup of Fessler and Eaton, as illustrated in Fig. 6.2.

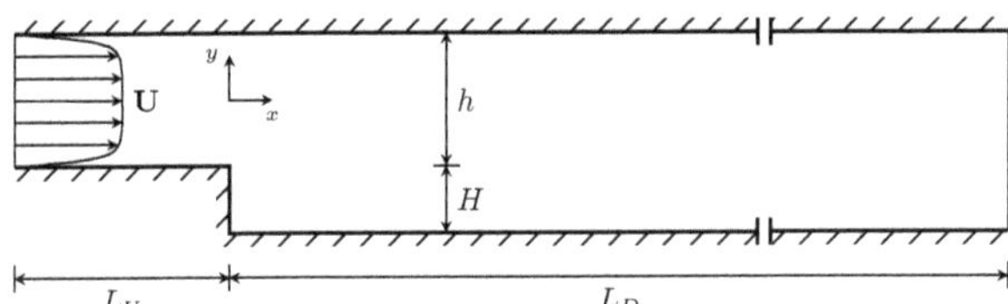

Fig. 6.2: Computational domain of the particle-laden turbulent backward-facing step flow, according to Fessler and Eaton [47].

The step and entrance channel height is $H = 0.0267\,\text{m}$ and $h = 0.04\,\text{m}$, which results in an expansion ratio of about $h/H = 5/3$. The length of the upstream and downstream channel is set to $L_U = 10h$ and $L_D = 35h$, while the channel span is $B = 4.28h$. The relatively long upstream channel is necessary to establish fully turbulent inlet conditions using a recycling method, which copies the (scaled) turbulent velocity from a plane downstream the channel entrance back onto the inlet (see section 4.4). Periodic boundary conditions were applied in spanwise direction, whereas no slip boundary conditions were set at the lower and upper channel walls. To ensure grid independence, three structured grids with 8.9×10^5, 2.9×10^6 and 7.0×10^6 cells were used. The boundary layer is directly resolved by placing the first grid point within the viscous sublayer ($y^+ \approx 1$) using a gentle grid stretching only in wall-normal direction. The Reynolds number based on the centerline velocity $u_0 = 10.5\,\mathrm{m/s}$ of the upstream channel (at $y/h = 0.5$) and the step height is $Re = u_0 H/\nu = 18{,}700$. Additionally, spherical monodisperse copper particles with a mass-averaged diameter of $D_p = 70\,\mu\mathrm{m}$ and a density of $\rho_p = 8{,}800\,\mathrm{kg/m^3}$ are randomly inserted at the flow inlet. The large-eddy Stokes number (based on a large-eddy passing frequency in the separated shear layer [47]) is $St = \rho_p D_p^2 u_0 / 10\mu_f 5H = 6.9$. According to Fessler and Eaton, the mass loading of the injected particulates is $\eta = \dot{m}_p/\dot{m}_f = 10\%$. The estimated volume fraction of the dispersed phase is $\phi_p < 10^{-3}$, which corresponds to a dilute flow and allows to neglect the inter-particle collisions (see section 3.1). Hence, only two-way coupling is considered during the simulations by applying a time-dependent volumetric source term, Eq. (3.19), to the momentum balance equation. In addition, no stochastic approach to model the influence of unresolved subgrid-scale velocity fluctuations on particle motions is applied (see section 3.5). All simulations were performed using second-order accurate approximations for the time (backward difference scheme) and spatial (central difference scheme) discretization.

6.2.2 Fluid statistics

Fig. 6.3 presents the distribution of the instantaneous u/u_0 and time-averaged streamwise velocity $\langle u/u_0 \rangle$ in the $x - y$ midplane of the domain, obtained for the medium grid resolution (i.e., 2.9×10^6 cells), which is proved to be sufficient for this flow configuration by a grid convergence study [67], and the DOE subgrid-scale model.

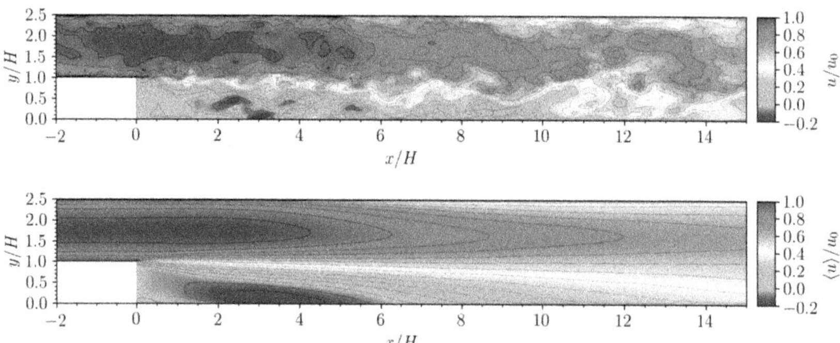

Fig. 6.3: Unladen turbulent backward-facing step flow: Distribution of the instantaneous streamwise velocity u/u_0 (top) and time-averaged streamwise velocity $\langle u \rangle/u_0$ (bottom) in the $x - y$ plane at $z/H = 0$ (midplane), obtained from LES (DOE).

The characteristic recirculation zone behind the backward-facing step, where backflow occurs, becomes clearly visible in the time-averaged distribution of the streamwise flow velocity, with the expected reattachment of the turbulent boundary layer at $x/H \approx 7$.

The instantaneous velocity field reveals detached shear layer directly behind the step at $x/H = 0$, which causes an enhancement of turbulence kinetic energy k behind the step all the way down to the outlet, as shown in Fig. 6.4.

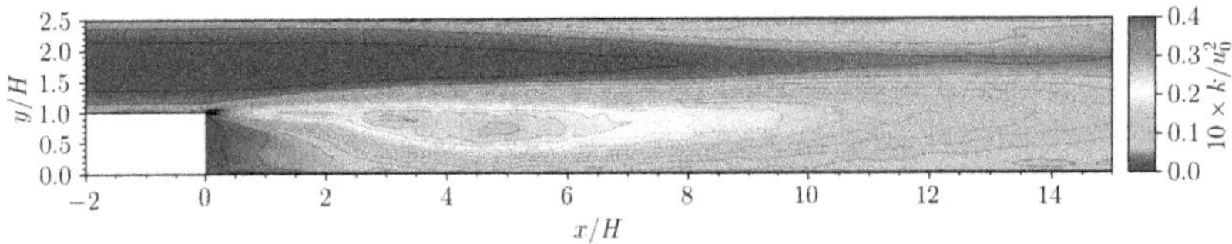

Fig. 6.4: Unladen turbulent backward-facing step flow: Distribution of the turbulence kinetic energy k/u_0^2 in the $x-y$ plane at $z/H = 0$ (midplane).

The calculated time-averaged streamwise velocity and Reynolds-stress profiles of the carrier flow are presented in Fig. 6.5, which are slightly amplified for the reasons of clarity and comprehensibility.

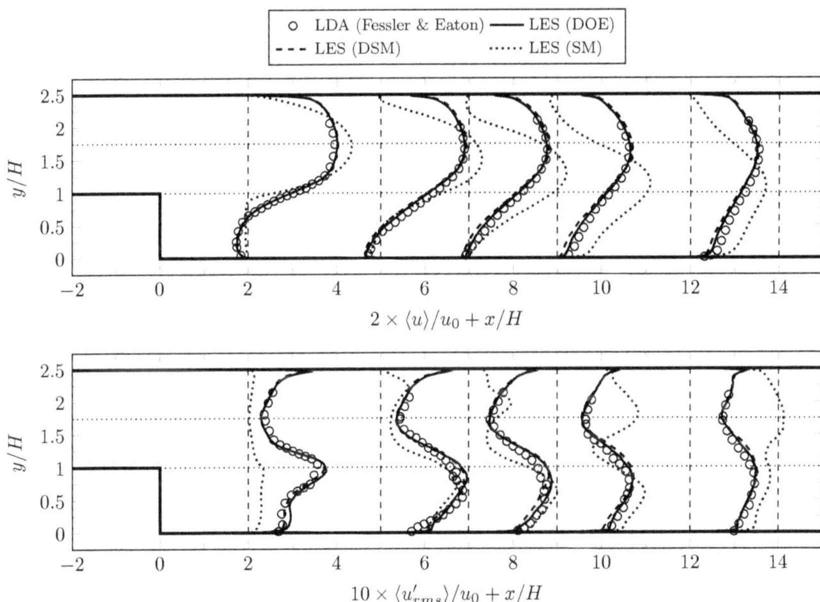

Fig. 6.5: Comparison of the time-averaged flow profiles of the carrier flow (continuous phase) for the streamwise velocity $\langle u \rangle / u_0$ (top)and streamwise velocity fluctuations $\langle u'_{rms} \rangle / u_0$ (bottom) behind the backward-facing step with the experimental LDA data of Fessler and Eaton [47] at $x/H = 2, 5, 7, 9, 12$.

The mean flow profiles $\langle u \rangle / u_0$ show again the characteristics of a typical BFS flow with a strong recirculation zone which arises immediately behind the step at $x/H = 0$ and an attachment of the turbulent boundary layer at approximately at $x/H = 7$. While the streamwise velocity profiles in case of the DSM and DOE model are very close to each other and to the experimental values, the SM model fails to predict the velocity field significantly. An explanation can be found by analyzing the streamwise velocity fluctuations $\langle u'_{rms} \rangle$, revealing strong velocity fluctuations (i.e., high turbulence kinetic energy k) within the shear layer, as getting visible in Fig. 6.4, which is captured well by the dynamic subgrid-scale models (DOE and DSM). For the SM model, the development of the shear layer is delayed, especially in close vicinity to the step ($x/H = 2$). The missing or retarded shear layer finally leads to the wrong calculation of the mean velocity field. It is supposed that the standard SM model fails to predict the BFS flow due to assumption of equilibrium between production and dissipation of small-scale turbulent scales (see section 2.4.3), which is not valid for separating and reattaching flows as the BFS flow.

6.2.3 Particle statistics

Fig. 6.6 presents the time-averaged particle velocity profiles $\langle u_p \rangle / u_0$ for the particle-laden BFS flow obtained from LES after an injection duration of 2 s in comparison to the measurements.

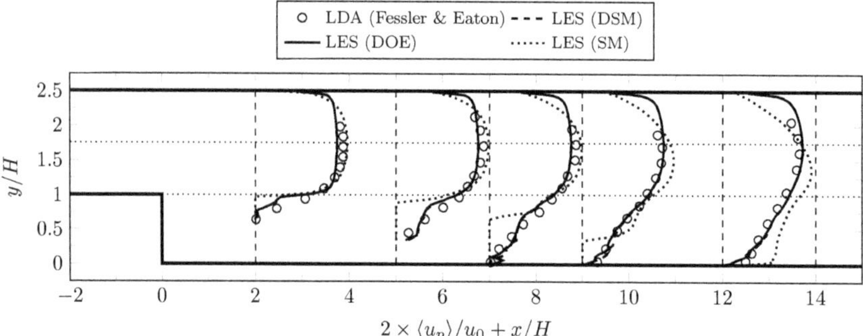

Fig. 6.6: Comparison of the time-averaged streamwise particle velocity profiles $\langle u_p \rangle / u_0$ (dispersed phase) behind the backward-facing step with the experimental LDA data of Fessler and Eaton [47] at $x/H = 2, 5, 7, 9, 12$.

It can be clearly seen, that the particles primarily follow the main fluid flow, which is expected for a relatively low particle Stokes number of $St = 6.9$. Thus, the magnitude and the shape of the mean particle velocity profiles are similar to the flow profiles of the continuous phase. The comparison between the numerical and experimental data reveals a minor underestimation of the particle velocities in the shear flow region in case of the DSM and DOE model, which disappears with an increasing distance to the step. An important secondary observation is, that the particle distribution or particle dispersion is correctly captured by the LES (DSM and DOE) without employing any additional stochastic tracking models, i.e., explicit modeling of sub-grid scale velocity fluctuations of the continuous phase (see section 3.5). The expansion of the particle cloud in all measuring points is in excellent agreement with the LDA measurements of Fessler and Eaton.

Concerning the presented results for the particle-laden turbulent BFS flow one can conclude that underlying LPT approach is capable of predicting the particle motion in shear-layer flows, including mean flow separation and reattachment, with a sufficient accuracy. Except in the case of the standard SM model, which is therefore not used for further investigations. Any essential influence of the unresolved sub-grid scales on the mean particle velocities was not observed. Moreover, it was possible to capture the turbulent particle dispersion without using any additional dispersion model.

6.3 Particle-laden flow in a simplified combustion chamber

The dispersed two-phase flow downstream a confined bluff-body (CBB) is often used for model validation (e.g., Apte et al. [7], Alletto and Breuer [3], Chrigui et al. [22]) and represents one of the simplest turbulent recirculating flows, representing a (cold) simplified combustion chamber without swirl. Although its geometrical simplicity, different features of this dispersed two-phase flow configuration, such as two recirculation zones with stagnation points (S_1 and S_2) and the complex particle dispersion due to an inner and outer shear layer, allow a challenging evaluation of the proposed Eulerian-Lagrangian approach with practical relevance.

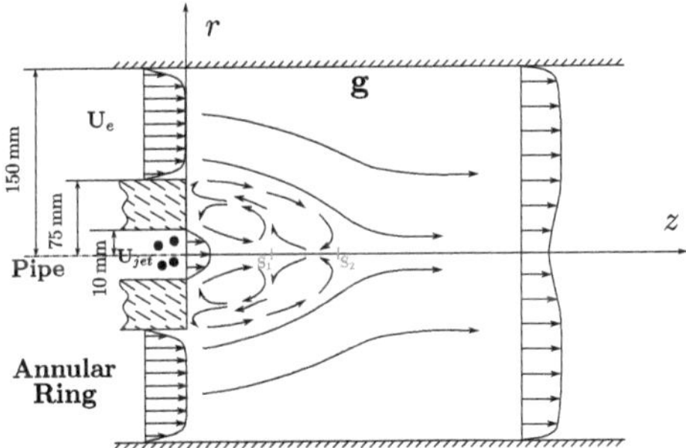

Fig. 6.7: Sketch of the computational domain for the (cold) particle-laden flow trough a simplified combustion chamber, according to Borée et al. [16].

6.3.1 Case description and numerical setup

The geometry of the chosen CBB configuration is illustrated in Fig. 6.7 and matches the original configuration proposed by Borée et al. [16], who investigated this bluff-body flow experimentally using comprehensive LDA measurements, in order to obtain detailed particle and fluid statistics for a wide range of particle sizes ($D_p = 20 - 100\,\mu\text{m}$) and two different mass loadings ($\eta = 22\%, 110\%$). A particle-laden air flow with a streamwise mean velocity of $u_{jet} = 3.01\,\text{m/s}$ enters the chamber through a circular pipe ($R_{pipe} = 10\,\text{mm}$) located on the chamber (symmetry) axis. The Reynolds number based on the mean flow velocity and the pipe radius is $Re = 2{,}006$. The injected particles have a diameter of $D_p = 50\,\mu\text{m}$ and density of $\rho_p = 2{,}470\,\text{kg/m}^3$ (i.e., $\rho_p/\rho_f \approx 2{,}100$), whereas the mass loading of the dispersed phase is set to $\eta = 22\%$. Clean air enters the chamber through an annular ring ($R_i/R_{pipe} = 7.5$, $R_a/R_{pipe} = 15$) with a mean flow velocity of $u_e/u_{jet} = 1.78$. The gravitational acceleration acts in the streamwise flow direction (i.e., $z-$ direction). Similar to the BFS flow, the turbulent inflow conditions are provided by a recycling method (see section 4.4) and no slip boundary conditions were set at the walls. To save CPU time, particles hitting the wall downstream of the CBB were removed from the domain. The domain is spatially discretized by wall-resolving O-type block-structured grids with about 2.2×10^6, 4.9×10^6 and 8.4×10^6 cells. All presented simulations were conducted using second-order accurate approximations for the time (backward difference scheme) and spatial (central difference scheme) discretization. The dynamic one equation eddy-viscosity model (DOE) is employed for closure.

6.3.2 Fluid statistics

In order to provide a first overview of the general flow structure, Fig. 6.8 shows the instantanous streamwise velocity field u_z/u_{jet} and the time-averaged streamwise velocity field $\langle u_z \rangle/u_{jet}$ including the streamlines of the averaged carrier flow.

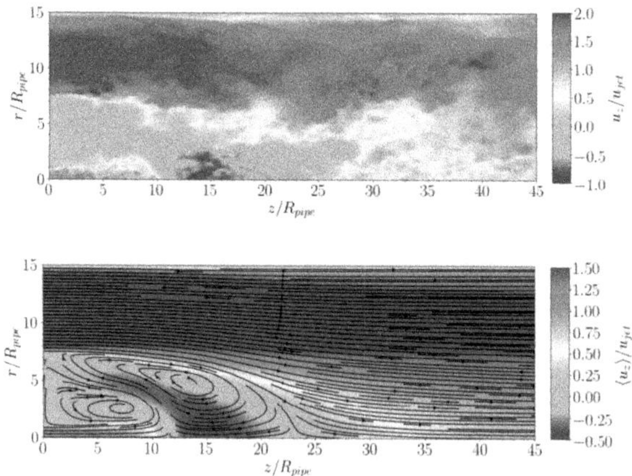

Fig. 6.8: Unladen turbulent flow in a simplified combustion chamber: Distribution of the instantaneous streamwise velocity u_z/u_{jet} (top) and time-averaged streamwise velocity $\langle u_z \rangle/u_{jet}$ including streamlines of the unladen carrier flow (bottom).

The distribution of the time-averaged streamwise velocity reveals the recirculation zones, indicated by negative flow velocities in the main flow direction, with stagnation points S_1 and S_2 (see Fig. 6.7) downstream of the CBB at $z/R_{pipe} \approx 12$ and $z/R_{pipe} \approx 23$. Within this flow configuration, areas with high fluctuating velocities are formed directly behind the first stagnation point and in the passage between the core and annular flows. The location of this stagnation points can be further specified by analyzing the time-averaged streamwise velocity $\langle u_z \rangle / u_{jet}$ of the carrier flow along the chamber symmetry axis (i.e., $r/R_{pipe} = 0$), as shown in Fig. 6.9. The stagnation points are located at the intersection of the velocity profiles with the dashed line at $\langle u_z \rangle / u_{jet} = 0$.

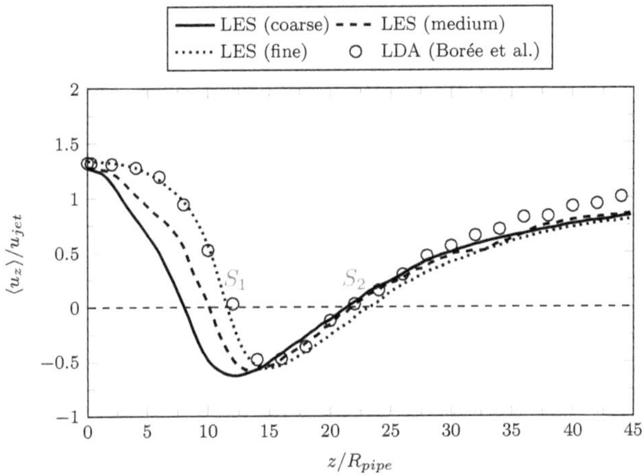

Fig. 6.9: Unladen turbulent flow in a simplified combustion chamber: Comparison of the time-averaged streamwise velocity $\langle u_z \rangle / u_{jet}$ of the carrier flow (continuous phase) along the chamber symmetry axis with the experimental LDA data of Borée et al. [16].

The prediction of the location for stagnation point S_1 is more sensible with regard to the grid resolution and is shifted towards the CBB, especially for the coarse mesh. However, the discrepancy between the LES results and the experimental data can be drastically reduced by increasing the mesh resolution. Surprisingly, the location of stagnation point S_2 as well as the increase of the of the streamwise velocity in the second part (downstream of S_2) is in good agreement with the LDA measurements, regardless of the grid resolution. Fig. 6.10 shows the time-averaged streamwise velocity $\langle u_z \rangle / u_{jet}$ and velocity fluctuation profiles $\langle u'_{z,rms} \rangle / u_{jet}$ of the carrier flow (continuous phase) along the z-axis compared to the LDA measurements. It can be highlighted that the results obtained from LES are accordance with the experiments, especially the predicted time-averaged velocity fields are in close agreement to the measured one. The streamwise velocity fluctuations profiles are slightly overestimated in the outer shear layer region (i.e., at $r/R_{pipe} = 7.5$), which cannot be improved by an increased grid resolution.

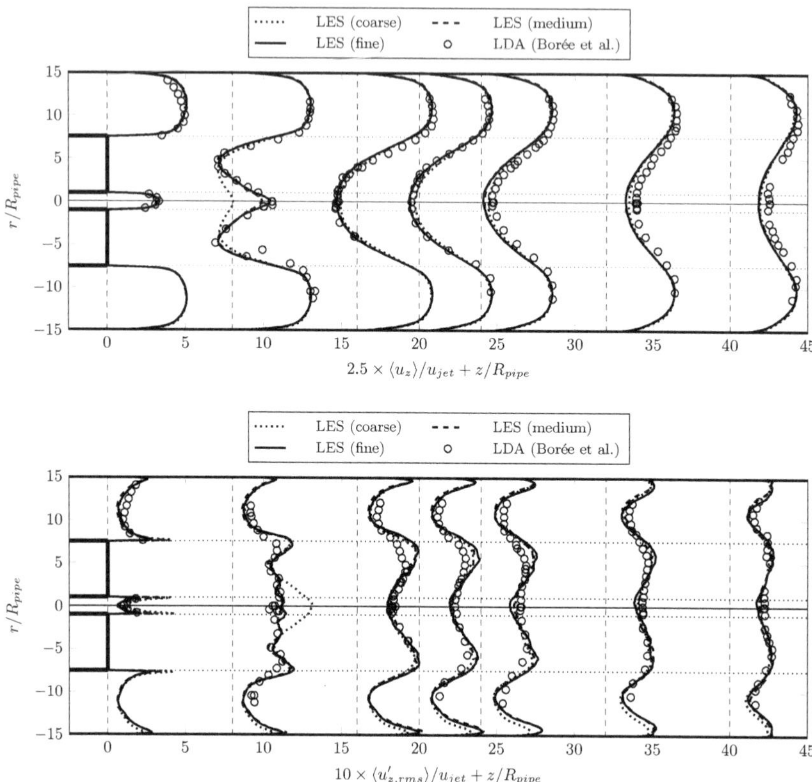

Fig. 6.10: Unladen turbulent flow in a simplified combustion chamber: time-averaged streamwise $\langle u_z \rangle / u_{jet}$ (top) and velocity fluctuation profiles $\langle u'_{z,rms} \rangle / u_{jet}$ (bottom) of the carrier flow (continuous phase) along the z-axis in comparison with the experimental data of Borée et al. [16].

6.3.3 Particle statistics

Based on the aforementioned LES results for the unladen flow downstream of the CBB, the medium grid resolution with about 4.9×10^6 cells is chosen for the Eulerian-Lagrangian LES of the corresponding particle-laden flow. In order to investigate the influence of the unresolved subgrid-scales of the continuous phase on the particle motion as well as the importance of inter-particle collision dispersed two-phase flows with a moderate mass loading, two- and four-way coupled simulations with and without explicit modeling of the particle dispersion were performed. To account for particle-particle collision, the soft-sphere model, as discussed in section 3.6.1, is applied.
Fig. 6.11 presents the time-averaged streamwise particle velocity $\langle u_{p,z} \rangle / u_{jet}$ and particle

velocity fluctuation profiles $\langle u'_{p,z,rms}\rangle/u_{jet}$ of dispersed phase along the z-axis. The abbreviation "PD" refers to explicit modeling of particle dispersion, according to section 3.5.

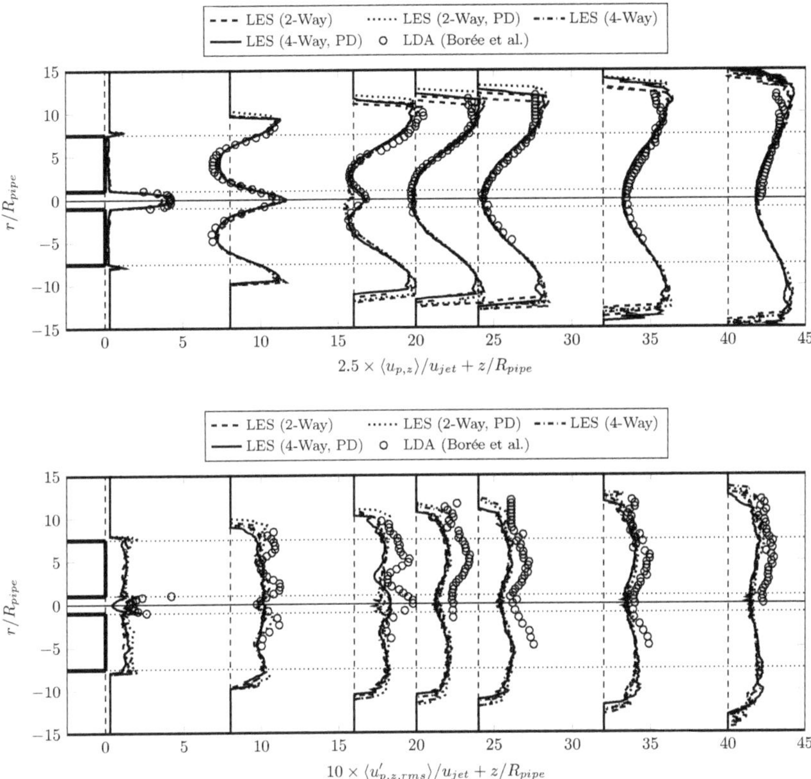

Fig. 6.11: Particle-laden turbulent flow in a simplified combustion chamber: time-averaged streamwise particle velocity $\langle u_{p,z}\rangle/u_{jet}$ (top) and particle velocity fluctuation profiles $\langle u'_{p,z,rms}\rangle/u_{jet}$ (bottom) of the dispersed phase along the z-axis in comparison with the experimental data of Borée et al. [16].

No remarkable influence of inter-particle collisions (i.e., four-way coupling) on the time-averaged streamwise particle velocity and particle velocity fluctuations is noticeable, which can be explained by the moderate mass loading and therefore the relatively low volume fraction of the dispersed phase. A similar observation can be done regarding the application of an explicit modeling approach for the particle dispersion. The mean particle velocity profiles are in excellent agreement to the LDA data, whereas slight deviations are discernible for the particle velocity fluctuations. However, the dispersion of the dispersed phase is very well predicted. From the experiments it must be expected that particle dispersion from

the inner shear layer at $r/R_{pipe} = 1.0$ through the outer shear layer at $r/R_{pipe} = 7.5$ into the annular jet region should occur, which is captured in the simulation, both qualitatively and quantitatively.
Despite of the minor discrepancies between the calculated and measured streamwise particle velocity fluctuations, the presented results underlines the high accuracy of the proposed Eulerian-Lagrangian approach, even for complex dispersed two-phase flows.

6.4 Particle-laden turbulent channel flow

The dispersed two-phase flows over a backward-facing step and through a simplified combustion chamber are dominated by specific features of the continuous phase such as recirculation zones and shear layers, whereas the influence of walls is secondary regarding the occurring particle dynamics. In contrast to this, in wall-dominated particle-laden flows (i.e., pipe and channel flows) different near-wall driving mechanisms are responsible for particle concentration build up in the near-wall accumulation region and for particle deposition in the deposition zone. These transport mechanisms are summarized in Fig. 6.12.

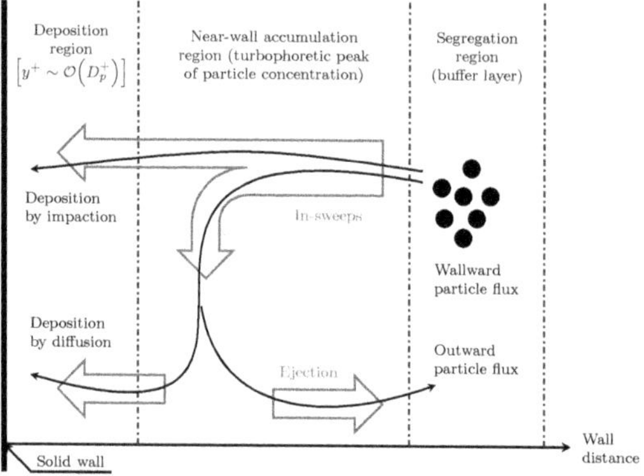

Fig. 6.12: Near-wall driving mechanisms, responsible for particle concentration build up in the near-wall accumulation region, according to Soldati and Marchioli [134].

According to Soldati and Marchioli [134], particles segregate and form coherent clusters in regions of the buffer layer (see also Fig. 6.19) where in-sweeps (i.e., local high momentum fluid flows in wall-direction within high velocity regions), generated by coherent low-speed streaks, can entrain them. Particles entrained in a sweep experience a net drift toward the near-wall accumulation zone, mainly due to turbophoresis[1] [98, 111], where the particle concentration reaches its maximum. Once the particles enter the accumulation

[1] Particle migration to the wall in turbulent boundary layers is referred to as turbophoretic drift. It is attributed to the non-homogeneous distribution of turbulent velocity fluctuations in the wall-normal direction (see e.g., Reeks [119]).

region, which is located in the viscous sublayer, they may either deposit at the wall or be reentrained toward the outer flow by ejections, generated by low-speed streaks in the near-wall region. It can be distinguished between two different main deposition mechanisms or scenarios. The first scenario is that particles that have acquired enough momentum may float through the accumulation region and deposit by impaction directly at the wall. The second one is that after a long residence time, particles can deposit under the action of turbulent fluctuations which are strictly zero only at the wall and, due to turbulence non-homogeneity, are always stronger in driving particles to the wall [134].
In order to investigate the features of wall-dominated flows and to ensure that all near-wall driving mechanisms are correctly captured in the following fouling simulation, a particle-laden turbulent channel flow is simulated using Eulerian-Lagrangian LES. The properties of the continuous or fluid phase as well as of the dispersed phase and the computational domain are the same as in the work of Marchioli et al. [100], who investigated this channel flow case by means of direct numerical simulations (DNS) for three different sets of spherical monodisperse particles and for Stokes numbers up to $St = 25$.

6.4.1 Case description and numerical setup

The computational domain of the particle-laden turbulent channel flow is shown in Fig. 6.13, with a size of $4\pi h \times 2h \times 2\pi h$ in x-, y- and z-direction and a half channel height of $h = 0.02$ m.

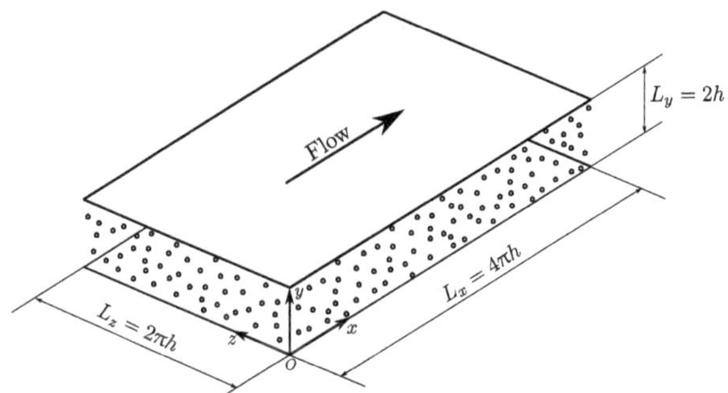

Fig. 6.13: Sketch of the computational domain for the particle-laden turbulent channel flow, according to Marchioli et al. [100].

The continuous phase, which carries the particles, is a turbulent gas flow. Air (assumed to be incompressible and Newtonian) is considered as working fluid with a density $\rho = 1.3\,\text{kg/m}^3$ and a kinematic viscosity $\nu = 1.57 \times 10^{-5}\,\text{m}^2/\text{s}$. According to Marchioli et al. [100], a nominal shear Reynolds number of $Re_\tau = u_\tau h/\nu = 150$, based on the shear velocity u_τ and the half of the channel height h, is chosen. The shear velocity is defined as $u_\tau = (\tau_w/\rho)^{1/2}$, where τ_w is the mean shear stress at the channel wall. The corresponding bulk Reynolds number is $Re_b = u_b h/\nu = 2{,}260$ based on the bulk velocity $u_b = 1.774\,\text{m/s}$.

For comparative purposes, all presented variables are in dimensionless form and expressed in inner coordinates (i.e., in wall units). Two types of spherical, rigid particles with an equal particle density $\rho_p = 1{,}000\,\mathrm{kg/m^3}$ but different particle diameters, namely $D_p = 20.4\,\mu\mathrm{m}$ and $D_p = 102.0\,\mu\mathrm{m}$, are considered in this study. The particle parameters including the Stokes number $St = \tau_p/\tau_f$, based on the particle relaxation time $\tau_p = \rho_p D_p^2/18\mu$ and the characteristic time scale of the flow $\tau_f = \nu/u_\tau^2$, are summarized in Tab. 6.1. To ensure representative sets of particle statistics, 10^5 particles are randomly placed in the channel domain with a fully developed carrier phase flow as initial conditions for the fluid, whereas the initial velocity of the particles is equal to the local fluid velocity. Due to the relatively low particle concentration, which allows one to consider dilute flow conditions, two-way phase coupling is applied (i.e., particle-particle interactions are neglected) and merely the drag is considered as acting particle force (for further details, see Marchioli et al. [100]). Once a statistically steady state is reached, averaging is conducted over a time interval of $85h/u_\tau$ (about 100 flow-trough times), followed by a combined streamwise and spanwise averaging over the complete channel.

Tab. 6.1: Particle parameters used in the Euler-Lagrangian LES, where the non-dimensional particle diameter D_p^+ is based on shear velocity $u_\tau = 0.11775\,\mathrm{m/s}$.

$St = \tau_p^+$	τ_p[s]	D_p^+	D_p [μm]
1	1.133×10^{-3}	0.153	20.4
25	2.832×10^{-2}	0.765	102.0

The computational domain is spatially discretized using three different mesh resolutions with $76 \times 46 \times 96$, $152 \times 46 \times 192$ and $152 \times 92 \times 192$ cells, resulting in a total cell number of 335,616, 1,342,464 and 2,684,928. Further parameters related to the spatial discretization are given in Tab. 6.2. Periodic boundary conditions are imposed on the continuous and dispersed phase in streamwise (x) and spanwise (z) direction, whereas no-slip conditions are applied for the parallel, flat, infinite channel walls. Additionally, the fluid flow is driven by a time-varying pressure drop $\partial p/\partial x$, to preserve a constant mean flow velocity and closure of the LES equations is provided by the DOE model.

Tab. 6.2: Different mesh resolutions used for the particle-laden turbulent channel flow at $Re_\tau = 150$.

Resolution	Re_τ	$N_x \times N_y \times N_z$	Δx^+	Δy^+_{min}	Δz^+
coarse	145.2	$76 \times 46 \times 96$	24.0	0.9	9.5
medium	146.1	$152 \times 46 \times 192$	12.1	0.9	4.8
fine	147.7	$152 \times 92 \times 192$	12.1	0.5	4.8

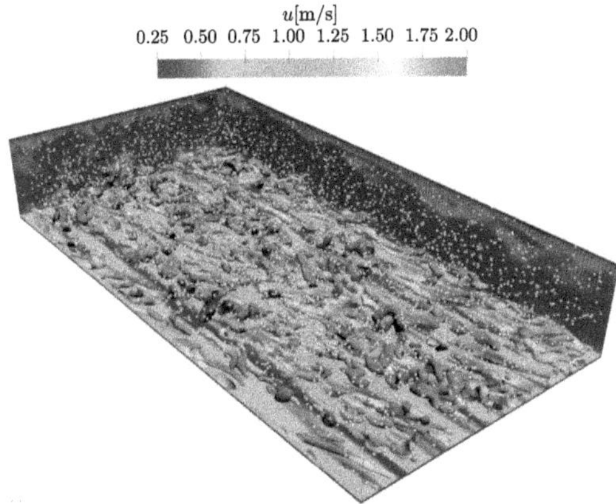

Fig. 6.14: Distribution of the spherical, monodisperse particles (only every 20th particle is displayed, with an 50 times enlarged diameter) and instantaneous streamwise fluid velocity u in the cross-sectional $x-z$ plane at $y^+ = 6.5$, including coherent near-wall structures ($\lambda_2 = 2.5 \times 10^3$).

6.4.2 Fluid statistics

In this section, the first and second order statistics (mean velocity and root mean square (rms) velocity fluctuations) of the clear fluid phase (without suspended particles), calculated by LES are presented and compared to the available DNS benchmark data. Fig. 6.15 shows the mean streamwise velocity $\langle u^+ \rangle$ obtained by the conducted LES, the DNS reference data as well as the analytical mean velocity profile given by the law of the wall, $u^+ = y^+$, and by the log law $u^+ = 2.5 \log(y^+) + 5.5$. The LES results for all three mesh resolutions matches the DNS and analytical profiles well, especially in the inner layer, which includes the viscous sublayer ($y^+ < 5$) and the buffer layer ($5 < y^+ < 30$). Slightly higher deviations become evident in the log-law region ($y^+ > 30$) for all LES, but this mismatch can be attenuated by a higher grid resolution. The rms velocity profiles $\langle u'^+_{rms} \rangle$, $\langle v'^+_{rms} \rangle$ and $\langle w'^+_{rms} \rangle$ are given and compared in Fig. 6.16. The LES results for all three cases, regardless of the mesh resolution, are in good accordance to the DNS data. Minor deviations are observed for the coarse numerical grid, where the characteristic peak of the streamwise velocity fluctuation at $y^+ \approx 12 - 14$, inside the buffer layer, is marginally overrated. The rms velocity profiles, obtained from LES, show a weak underestimation in all cases near the centerline, where the flow conditions are close to isotropic and homogeneous, but are still in line with the DNS data. Additionally, Fig. 6.17 presents the skewness S and flatness F (third- and fourth-order statistical moment, see section A.1) of the velocity components for all mesh resolutions. The skewness profile of the wall-normal velocity provides direct information about the turbophoretic drift (see e.g., [119, 133]) within the near wall region ($y^+ \leq 30$), which drives the particles towards the channel wall (negative y-direction). This turbulence induced drift can significantly influence the accumulation and deposition

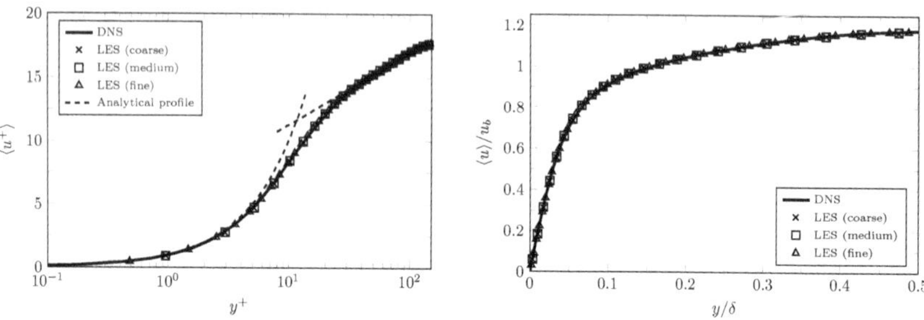

Fig. 6.15: Mean streamwise fluid velocity profile $\langle u^+ \rangle$ in inner (left) and outer coordinates (right).

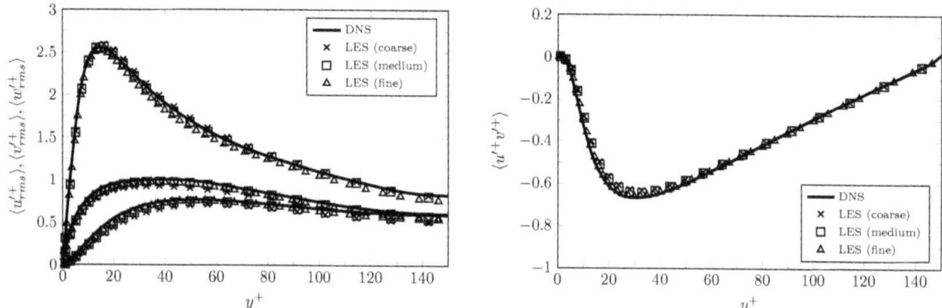

Fig. 6.16: Rms fluid velocity profiles $\langle u'^+_{rms} \rangle$, $\langle v'^+_{rms} \rangle$, $\langle w'^+_{rms} \rangle$ (left) and Reynolds stress component $\langle u'^+ v'^+ \rangle$ (right).

of particles and must be captured in the LES for reasonable results. In principle, all conducted LES show this drift, although with different accuracy. It has to be pointed out that a crucial damping of the velocity fluctuations in wall-normal direction, as it may occur due to LES filtering [12], was not observed. For the medium and fine mesh resolution, the skewness for all fluid velocity fluctuations is in excellent agreement with the DNS data. The reasonable accurate description of the flow field shows, that even the coarse mesh resolution already agrees with the well-resolved LES, in which the SGS modeling error for the fluid is negligible. However, based on this findings and to ensure a high accuracy, the medium mesh resolution is used in the following investigation of the particle statistics and particle distribution in dependency of the Stokes number.

6.4.3 Particle statistics and distribution

The following passage presents the mean streamwise particle velocity and the rms particle velocity fluctuations, obtained by Eulerian-Lagrangian LES, and highlights the influence of the particle size or Stokes number St, respectively, on the first and second order particle statistics. Fig. 6.18 shows the mean streamwise particle velocity $\langle u_p^+ \rangle$ calculated using

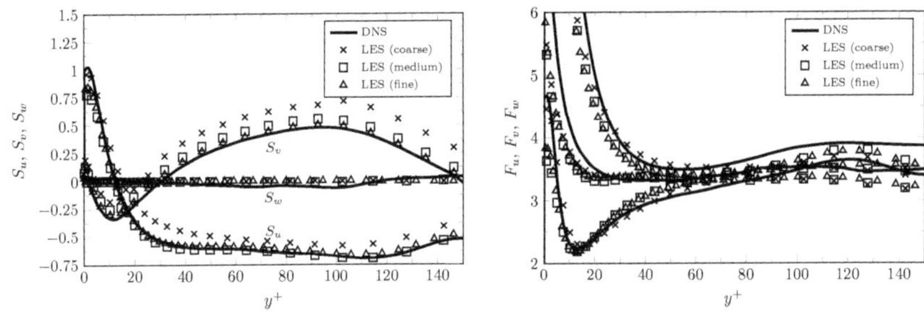

Fig. 6.17: Fluid velocity skewness factors S_u, S_v, S_w (left) and fluid velocity flatness factors F_u, F_v, F_w (rights).

LES for small and large particles, resulting in Stokes numbers of $St = 1$ and 25.

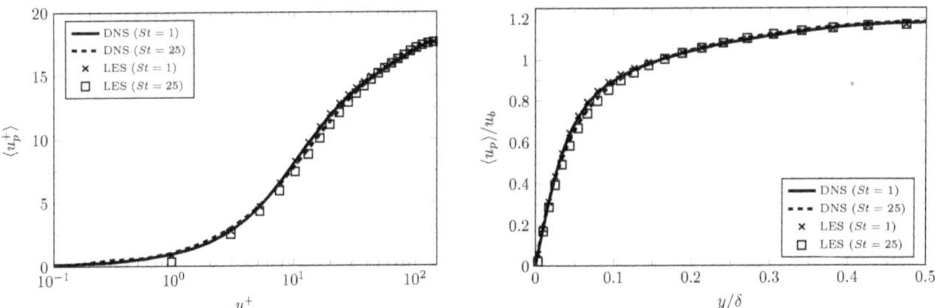

Fig. 6.18: Mean streamwise particle velocity profile $\langle u_p^+ \rangle$ in inner (left) and outer coordinates (right) and depending on the Stokes number St.

The calculated streamwise particle velocity profiles, for both particle sizes, are in excellent agreement with the DNS data. Especially the viscous sublayer and buffer region are well captured using Eulerian-Lagrangian LES, whereas larger deviations are visible in the log-law region of the flow, resulting from an overprediction of the carrier phase velocity. However, the Eulerian-Lagrangian LES is able to capture even very fine differences in the particle velocity profiles caused by the particle size or Stokes number, respectively, for example the lower velocities for the large particles ($St = 25$) in the beginning of the log-law region $\left(80 < y^+ < 110\right)$. Additionally, the segregation region within the buffer layer, where particles segregate and form coherent clusters, is captured well as shown in Fig. 6.19. A similar performance of the LES is obtained for the particle velocity fluctuations, illustrated in Fig. 6.20. The calculated rms fluctuations are clearly in line with the DNS data and show the expected behavior for both Stokes numbers. Due to the small size and inertia of the particles for $St = 1$, they follow sudden velocity changes in the carrier fluid instantly. Hence, the rms velocity fluctuations of the fluid phase (see Fig. 6.16) and the dispersed phase are equal for all directions. In case of the large particles, the particle velocity

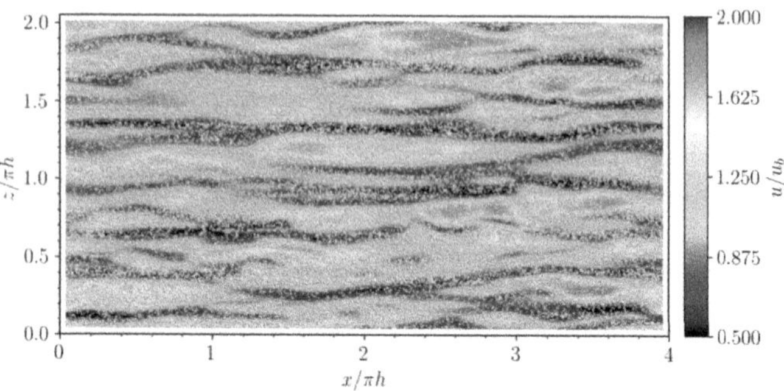

Fig. 6.19: Particle-laden turbulent channel flow: normalized instantaneous streamwise velocity u/u_b and distribution of the spherical particles ($St = 25$) in the cross-sectional $x - z$ plane at $y^+ \approx 10.5$ within the buffer layer.

fluctuations in the wall normal and streamwise direction are clearly damped or filtered by the high particle inertia. On the other hand, the streamwise rms particle fluctuations within the near-wall region are distinctly amplified, compared to the clean fluid flow and the particle-laden flow including small particles ($St = 1$). Accordingly, larger particles enhance the mean particle turbulence kinetic energy $\langle k_p^+ \rangle$, shown in Fig. 6.20, within the buffer layer and damp it near to the centerline (log-law region), where the turbulence is almost isotropic, without dominant velocity fluctuations. These results are in line with observations of Marchioli et al. [100, 99]. Since the precise calculation of particle

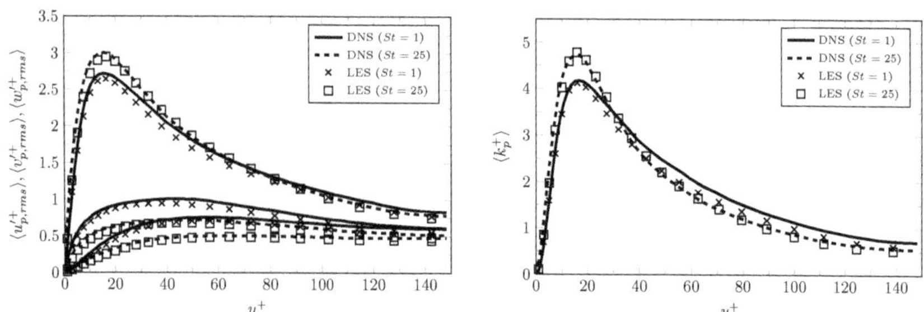

Fig. 6.20: Rms particle velocity profiles $\langle u'^+_{p,rms} \rangle$, $\langle v'^+_{p,rms} \rangle$, $\langle w'^+_{p,rms} \rangle$ (left) and particle turbulence kinetic energy $\langle k_p^+ \rangle$ (right) and depending on the Stokes number St.

distributions is of crucial importance for simulations of wall-bounded flows, especially for the physical prediction of particle depositions, the general behavior of smaller and larger particles in the near-wall region is further investigated. Fig. 6.21 presents the normalized mean particle concentration profiles $\langle C \rangle / C_0$ in wall units, showing a strong accumulation of particles at the channel walls, regardless of the particle size, which is decaying rapidly while passing the buffer region of the wall boundary layer. Moreover, a higher Stokes number leads to a increased particle concentration within the near wall region $\left(y^+ < 30\right)$ and a lower particle concentration within the remaining flow. For the small particles, the normalized particle concentration $\langle C\rangle/C_0$ in the viscous sublayer is about 2.3 and falls down to 1, while the particle concentration for the larger particles is significantly higher in the near-wall region and drops below 0.3 in the core flow. Thus, the majority of the large particles is bounded in the viscous wall region and cannot leave the inner boundary layer, even without the consideration of gravity. These findings are approved by numerical studies of Marchioli et al. [100] and Zhao et al. [171].

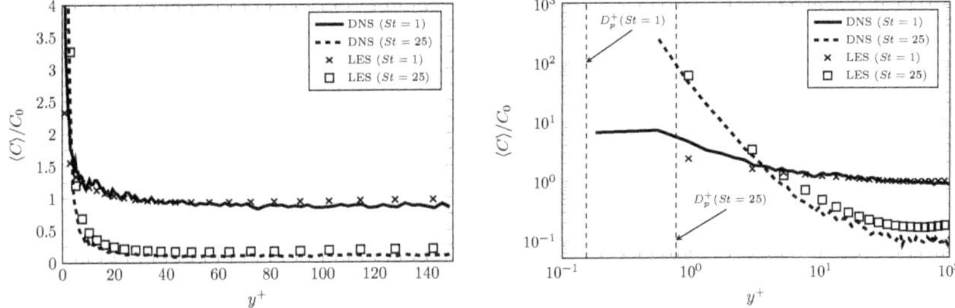

Fig. 6.21: Normalized mean particle concentration profiles $\langle C\rangle/C_0$ in dependency of the Stokes number St. The vertical dashed lines provide a visual indication of the size of the particle diameter.

6.4.4 Particle deposition

The implemented deposition model (see section 5.1.1) is validated using the particle deposition rate or particle deposition velocity for a horizontal channel (i.e., gravity acts perpendicular to the flow direction). Therefore, a time-averaged particle deposition rate V_d, which is defined as the ratio of the particle flux at the deposition surface to the particle concentration, is estimated as follows [161]:

$$V_d = \frac{N_d/(A\cdot t)}{N/V}, \tag{6.4}$$

where N_d is the number of deposited particles during the time period t. A is the area of deposition, N is the number of all particles within the turbulent channel flow and V is the volume of the domain. Fig. 6.22 shows the particle deposition rates for different particle relaxation times obtained from Eulerian-Lagrangian LES for the horizontal turbulent channel flow. The deposition rate is normalized by the mean friction velocity $\langle u_\tau\rangle$:

$$V_d^+ = \frac{V_d}{\langle u_\tau\rangle}, \tag{6.5}$$

whereas the dimensionless particle relaxation time is defined by comparing the particle relaxation time, Eq. (3.10), to the timescale associated with the near-wall turbulent eddies or the average lifetime of the smallest eddies near the wall [170]:

$$\tau_p^+ = \frac{\tau_p}{\tau_e} = \frac{\rho_p D_p^2 u_\tau^2}{18\mu_f \nu_f}. \tag{6.6}$$

The comparison of our LES results with the experimental measurements performed by Sehmel [128] and the DNS-Lagrangian data of Zhang and Ahmadi [169] reveals a similar trend of the particle deposition rate and a high coincidence. Discernible deviations are visible merely in the range of low ($\tau_p^+ < 1$, very fine particles) and higher ($\tau_p^+ > 50$, large particles) particle relaxation times, but the overall agreement is proven to be satisfactory. Thus, the implemented deposition model is capable of describing the particle deposition with a good accuracy for a wide range of particle relaxation times.

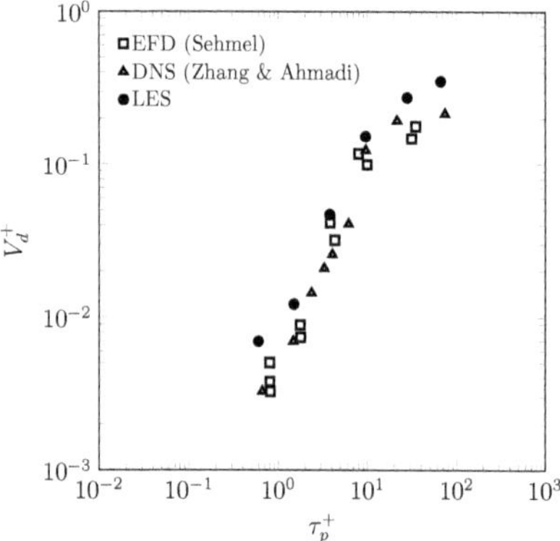

Fig. 6.22: Particle deposition rates obtained from Eulerian-Lagrangian LES compared with the experimental measurements proposed by Sehmel [128] and DNS data of Zhang and Ahmadi [169].

The presented fluid and particle statistics together with the particle deposition rates for a turbulent particle-laden channel flow underlines the great potential of the Eulerian-Lagrangian LES for the prediction of wall-dominated dispersed multiphase flows. Thus, this approach is proved to be suitable for the following fouling simulations, even though special attention needs to be paid to a sufficient high mesh resolution, to overcome an inaccurate prediction of particle velocity fluctuations as well as particle preferential segregation and accumulation at the wall due to an insufficient LES filter width (cut-off length) [99]. A comprehensive overview about the lack of any closure to account for the effect of the SGS fluid velocity fluctuations in the equation of particle motion is given by Marchioli [96].

7 Particulate fouling on dimpled heat transfer surfaces

This chapter is focused on numerical investigations of particulate fouling on different types of dimpled heat surfaces, namely the single spherical dimple and spherical dimples in a staggered arrangement (or dimple package). In the first part, the interaction between local flow structures, induced by a single spherical dimple, and the thermo-hydraulic performance as well as the formation of fouling deposits is investigated. The numerical results are than compared to experimental results provided by the Institute for Chemical and Thermal Process Engineering (ICTV) at the Technische Universität Braunschweig. The second part contains the prediction of particulate fouling for spherical dimples in a staggered arrangement. Since such dimple packages are employed for a broad range of industrial applications, this study will expand the fouling investigations for a pure academic test case, the isolated spherical dimple, to a case of application with a higher industrial relevance. The majority of the presented work was originally published in Kasper et al. [67, 66, 68] and Deponte et al. [34].

7.1 Particulate fouling on a single spherical dimple

Based on various experimental and numerical investigations of dimples regarding heat transfer and pressure loss (e.g., Terekhov et al. [143], Mahmood et al. [95], Ligrani et al. [85, 84], Elyyan et al. [42], Isaev et al. [62], Turnow et al. [148, 149, 146]), a smooth narrow channel with an isolated spherical dimple has been chosen to investigate the interaction between local flow structures and particulate fouling numerically, considering two dimple depth-to-dimple diameter ratios of $t/D = 0.26$ and $t/D = 0.35$, respectively, since these t/D-ratios are known for its superior thermo-hydraulic performance. For comparative purposes, the results for the spherical dimple are compared to these obtained for a square cavity with an equivalent cavity depth-to-cavity side length ratio.

7.1.1 Case description and numerical setup

The computational domain for the turbulent channel flow over a single spherical dimple and a square cavity is shown in Fig. 7.1. The origin of the coordinate system is located in the center of the spherical dimple or square cavity and is projected onto the lower wall plane, therefore the lower wall is located at $y/H = 0$. The length of the channel

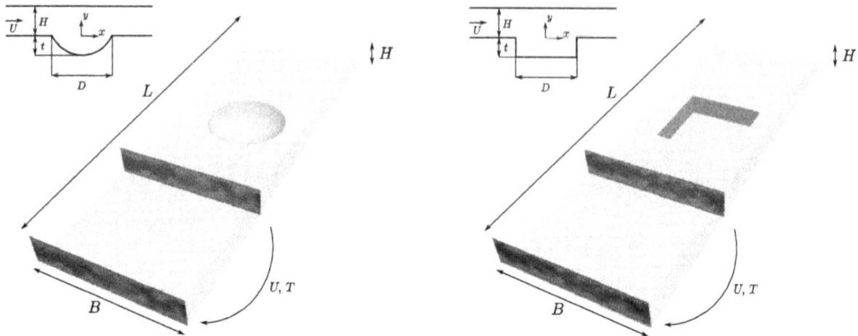

Fig. 7.1: Computational domain for a turbulent channel flow over a single spherical dimple and a square cavity.

is $L = 230\,\mathrm{mm}$, while channel height H and channel width B are set to $H = 15\,\mathrm{mm}$ and $B = 80\,\mathrm{mm}$. A dimple print diameter of $D = 46\,\mathrm{mm}$ as well as a dimple depth of $t = 12\,\mathrm{mm}$ and $t = 16\,\mathrm{mm}$ are chosen, while the side length of the square cavity equals the dimple print diameter D and the cavity depth is likewise set to $t = 12\,\mathrm{mm}$ and $t = 16\,\mathrm{mm}$, respectively. Periodic boundary conditions were applied in the spanwise direction, whereas no slip conditions were set at the lower and upper channel walls, wall-roughness effects are neglected. Turbulent inlet conditions were produced using a recycling method, which copies the turbulent velocity and temperature field from a plane downstream the channel entrance back onto the inlet and ensures a divergence free velocity field (see section 4.4). The normalized temperature $T^+ = (T - T_\infty) / (T_w - T_\infty)$ is used, where a constant $T^+ = 1$ is assumed at the lower wall. The molecular Prandtl number Pr was set to 0.71 in all simulations, whereas the turbulent Prandtl number Pr_t is 0.9. The Reynolds number based on the average bulk velocity $u_b = 0.91\,\mathrm{m/s}$, the kinematic viscosity $\nu = 1 \times 10^{-6}\,\mathrm{m^2/s}$ and the dimple print diameter D is $Re_D = 42{,}000$. The applied boundary conditions are summarized in Tab. 7.1. Additionally, the unresolved subgrid scales are modeled using the DOE model (see also section A.2.2).

Tab. 7.1: Boundary conditions applied for the LES of the turbulent channel flow over a spherical dimple and a square cavity.

Flow variable	Inlet	Outlet	Lower wall	Upper wall	Sides
Velocity	recycling	$\partial\overline{\mathbf{u}}/\partial\mathbf{n} = \mathbf{0}$	$\overline{\mathbf{u}} = \mathbf{0}$	$\overline{\mathbf{u}} = \mathbf{0}$	periodic
Pressure	$\partial\overline{p}/\partial\mathbf{n} = 0$	$\overline{p} = 0$	$\partial\overline{p}/\partial\mathbf{n} = 0$	$\partial\overline{p}/\partial\mathbf{n} = 0$	periodic
Temperature	recycling	$\partial\overline{T^+}/\partial\mathbf{n} = 0$	$\overline{T^+} = 1$	$\overline{T^+} = 0$	periodic

To assure grid independence of the obtained results, a series of calculations on different grid resolutions was carried out (see also section A.2.1). Block-structured curvilinear grids consisting of about 7.8×10^5, 1.6×10^6 and 3.4×10^6 cells were used, as presented in Tab. 7.2. In spanwise (z) and streamwise directions (x), an equidistant grid spacing is applied, whereas in wall-normal direction (y) a homogeneous grid stretching is used to place the first grid node inside of the viscous sublayer at $y^+ \approx 1$.

Tab. 7.2: Different mesh resolutions used for the LES of the turbulent channel flow over a single spherical dimple.

Mesh	Number of cells N
M1	781,632
M2	1,637,433
M3	3,384,834

Spherical monodisperse silicon dioxide particles (SiO_2) with a diameter of $D_p = 20\,\mu m$, typical for fouling related problems, and a density ratio of $\rho_p/\rho_f = 2.5$ are randomly injected at the domain inlet during the fouling simulations. The thermo-physical properties of these particles are given in Tab. 7.3. Considering earlier experimental fouling investigations on heat exchangers by Blöchl and Müller-Steinhagen [13], a particle mass loading of $\eta = \dot{m}_p/\dot{m}_f = 0.1\%$ and $\eta = 0.2\%$ is chosen to ensure an asymptotic fouling layer growth and a steady-state conditions within a few minutes of physical real time. The motion of the suspended particles is predicted considering drag, lift, gravity and buoyancy, added mass as well as pressure gradient force. Due to the low estimated volume fraction of the dispersed phase ($\alpha_p < 10^{-3}$), the negligence of inter-particle collisions is assumed. Thus, only two-way phase coupling is applied.

Tab. 7.3: Thermo-physical properties of the injected foulant particles.

Physical property	Value
Diameter	$D_p = 20\,\mu m$
Density	$\rho_p = 2{,}500\,\mathrm{kg/m^3}$
Thermal conductivity	$k_p = 1.0\,\mathrm{W/(m\,K)}$
Specific heat capacity	$c_p = 1.052\,\mathrm{kJ/(kg\,K)}$

7.1.2 Flow structures

Despite their geometrical simplicity, dimples induce much more complex flow phenomena compared to those over a flat surface (i.e., plane channel flow), which are still subject of current experimental and numerical investigations. In order to provide a first overview of the these flow structures, Fig. 7.2 shows the distribution of the instantaneous streamswise flow velocity u/u_b, normalized by the average bulk velocity u_b, in the midplane of the channel flow over a single spherical dimple with a dimple depth-to-dimple diameter ratio of $t/D = 0.26$.

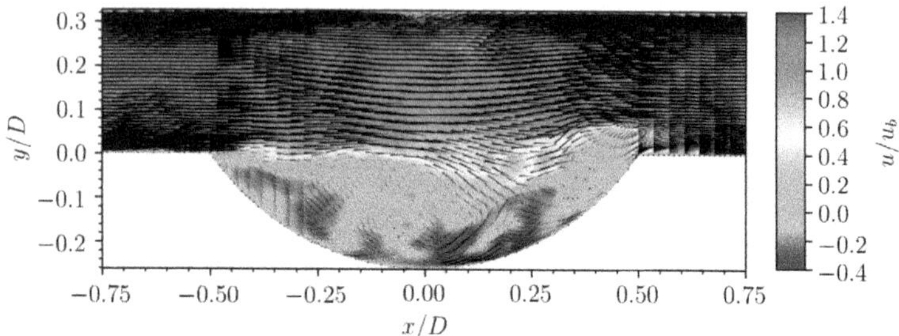

Fig. 7.2: Turbulent channel flow over a single spherical dimple with a dimple depth-to-dimple diameter ratio of $t/D = 0.26$ at $Re_D = 42{,}000$: Distribution of the instantaneous streamswise flow velocity u/u_b in the $x - y$ plane at $z/D = 0$ (midplane).

The distribution of the streamwise velocity u/u_b reveals essential flow features induced by dimpled surfaces. One of these characteristic flow phenomena is the generation of unsteady *Kelvin-Helmholtz* instabilities in the shear layer between the high-speed mainstream flow and the low-speed recirculating flow inside the dimple, which shed periodically along the interface between the two flow streams. It should be mentioned that the *Tollmien–Schlichting* instability may also contribute to the generation of unsteady vortices structures in the boundary layer flow over the dimpled surface [172]. The generation and periodic shedding of these transient Kelvin-Helmholtz vortices cause intensive turbulent mixing in the boundary layer over the dimple by forcing the entrainment of high-speed mainstream flow from the outer region into the near-wall region and promoting the ejection of the low-speed recirculating flow out of the dimple, which greatly enhances the heat transfer process between the high-speed coolant flow and the near-wall hot flow, as reported by Ligrani et al. [85, 84]. Another important flow phenomenon is the so-called *upwash flow* in the near of the back rim or trailing edge of the dimple, generated by the high-speed mainstream flow which is slightly shifted downward and impinge onto the back of the dimple. While the region near the front rim or front edge of the dimple is characterized by flow separation and thus reduced heat transfer and a low pressure region, causing the downward shift of mainstream flow into the dimple cavity, the impingement of the high-speed mainstream flow onto the back rim results in a high pressure region and an increased heat transfer over the back portion of the dimple. The strong upwash flow near the back rim also significantly enhance the turbulent mixing between the low-momentum flow near the wall and the high-momentum core flow (or freestream flow) away from the wall, resulting again in an augmented heat transfer process in the downstream region of the dimple. This upwash flow enhancing turbulent mixing in the boundary layer flow over the dimpled surface can be seen clearly in the distribution of the normalized turbulence kinetic energy k/u_b^2 sampled in the $x-z$ plane at $y/D = 0.08$, given in Fig. 7.3.

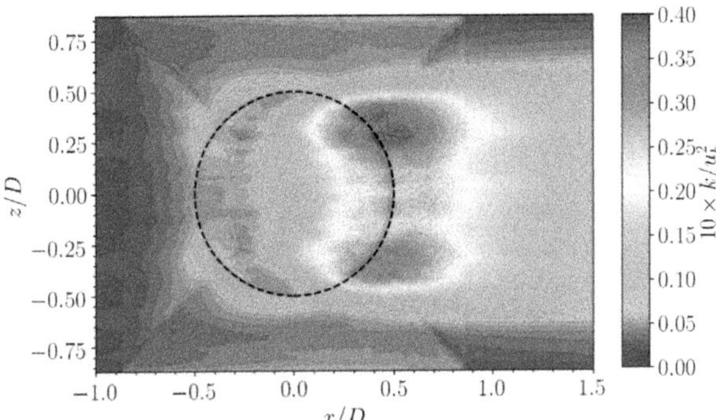

Fig. 7.3: Turbulent channel flow over a single spherical dimple with a dimple depth-to-dimple diameter ratio of $t/D = 0.26$ at $Re_D = 42{,}000$: Distribution of the turbulence kinetic energy k/u_b^2 in the $x-z$ plane at $y/D = 0.08$.

The aforementioned flow phenomena are schematically illustrated in Fig. 7.4. It should be noted that the intensity and direction of the upwash flow is primarily controlled by the dimple depth-to-dimple diameter ratio t/D as well as the Reynolds number Re_D based in the dimple diameter [172], which is further addressed in section 7.1.3.

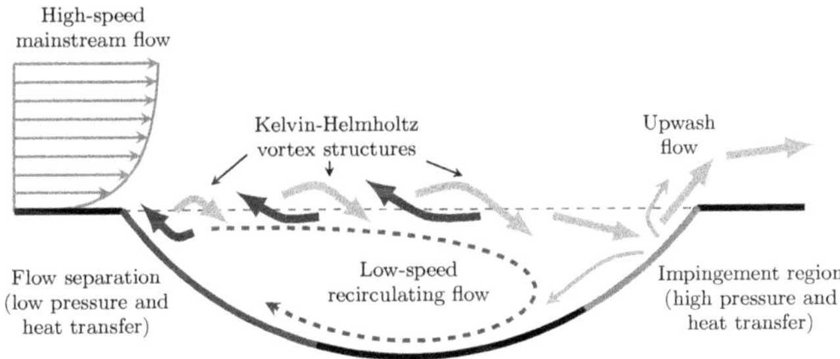

Fig. 7.4: Schematic of the flow phenomena induced by a spherical dimple.

Probably the most important flow feature generated by the investigated spherical dimples, regarding their fouling-mitigation potential (see section. 7.1.4), can be seen in the phase-averaged streamline pattern, given in Fig. 7.5. These streamlines reveal that the low-speed recirculating flow forms an unsteady asymmetrical mono-core vortex structure inside the

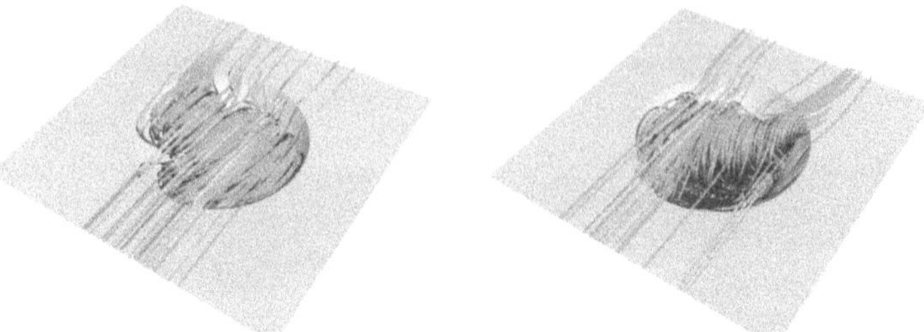

Fig. 7.5: Different orientations of the oscillating mono-core vortex structures (±45°) inside the spherical dimple for $Re_D = 42{,}000$ and a dimple depth-to-dimple diameter ratio of $t/D = 0.26$.

dimple, which alters its orientation arbitrarily from −45° to +45° with respect to the main flow direction and leaves the dimple cavity as ejecting jet at the back portion of the dimple. It is assumed that this oscillating jet promotes a self-cleaning process inside and in the vicinity of the dimple, which is further analyzed in section 7.1.4. The vortex switching is reflected in the time history of the instantaneous streamwise velocity u/u_b, obtained from LES and sampled in two symmetric points located at $x/D = 0$, $y/D = 0$, and $z/D = \pm 0.217$ (i.e., within the interface or dividing surface between channel and dimple), as presented in Fig. 7.6.

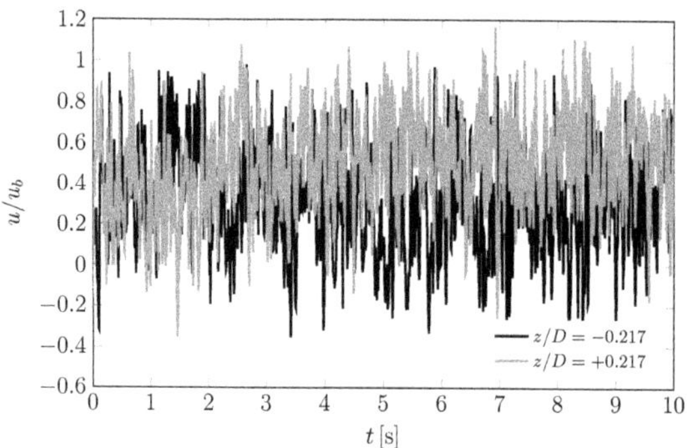

Fig. 7.6: Time history of the instantaneous streamwise velocity u/u_b sampled in two symmetric points at $x/D = 0$, $y/D = 0$, and $z/D = \pm 0.217$ for $Re_D = 42{,}000$ and $t/D = 0.26$.

In order to get a deeper insight into the vortex dynamics, the (average) switching frequency of the mono-core vortex structure is determined using Welch's method [163] by placing two additional sample points at $x/D = 0.217$, $y/D = 0$, and $z/D = \pm 0.326$ (i.e., within the interface between channel and dimple cavity). In these sample points, the instantaneous streamwise velocity was recorded for a time interval of $t = 30\,\text{s}$ with a sample interval of $t_s = 5 \times 10^{-4}\,\text{s}$, resulting in a corresponding sample rate of $f_s = 2\,\text{kHz}$. Application of Welch's method on the gathered velocity data leads to the temporal frequency and energy spectrum shown in Fig. 7.7.

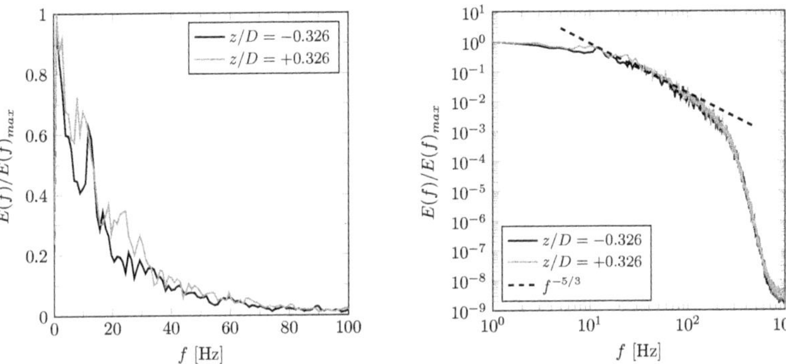

Fig. 7.7: Normalized temporal frequency (left) and energy spectrum (right) of the instantaneous streamwise velocity u sampled in two symmetric points at $x/D = 0.217$, $y/D = 0$, and $z/D = \pm 0.326$ for $Re_D = 42{,}000$ and $t/D = 0.26$.

The frequency spectrum reveals strong peaks at $f \approx 0.2\,\text{Hz}$, representing the switching frequency of the oscillating mono-core vortex structure inside the dimple. Thus, vortex switching appears approximately every 5 s in the average, which results in a minimum micro fouling time interval of $\delta t_f \approx 10\,\text{s}$ within the multiscale modeling approach presented in section 5.2.1, in order to capture a possible self-cleaning effect of oscillating ejecting jet generated by the dimple. This long period oscillations frequency is in line with experiments of Terekhov et al. [143] and numerical investigations of Turnow et al. [146]. In addition, the temporal energy spectrum given in Fig. 7.7 reproduces the inertial subrange with a $f^{-5/3}$ slope which underlines the accuracy of the conducted simulations. Furthermore, it is worth to mention that different studies (see e.g., Ligrani et al. [85]) proved that the appearance of vortex oscillations as well as the switching or shedding frequency depends merely on the dimple depth-to-dimple diameter ratio t/D and the Reynolds number Re_D (i.e., based on the dimple diameter D).

The numerical results for the unladen turbulent channel flow over a single spherical dimple are validated using the experimental data published by Terekhov et al. [143] and Turnow et al. [146] for Reynolds number $\text{Re}_D = 40{,}000$. Fig. 7.8 shows the profiles of the streamwise velocity $\langle u \rangle / u_0$ and streamwise velocity fluctuations $\langle u'_{rms} \rangle / u_0$ obtained from LES and URANS ($k - \omega$ SST model [103]) for three different grid resolutions and from LDA measurements in the midplane of the channel ($z/H = 0$) at $x/D = 0, \pm\,0.217$. A satisfactory overall agreement of the calculated and measured time-averaged streamwise velocity profiles $\langle u \rangle / u_0$ has been obtained for all three grid resolutions (see Tab. 7.2) and streamwise locations x/D. The velocity profiles from LES match well with the

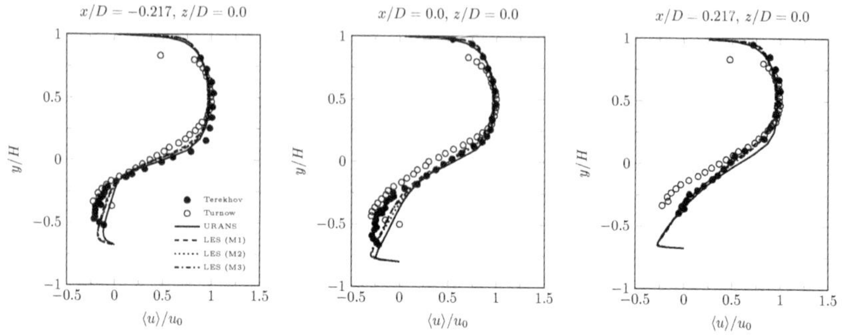

Fig. 7.8: Time-averaged streamwise velocity profiles $\langle u\rangle/u_0$ (normalized by the center line velocity u_0) obtained from LES and RANS for different streamwise positions $x/D = 0, \pm0.217$ in the midplane of the channel ($z/D = 0$) in comparison with LDA measurements of Terekhov et al. [143] and Turnow et al. [146].

measurements in the center of the channel, where the maximum flow velocity occurs and even in the upper near-wall region. Slight deviations from the measured profiles can be registered for all grid resolutions inside of the spherical dimple within the low-speed recirculating flow. However, since the LES results are in good agreement in the remaining regions, the most evident reason for the discrepancy are possible LDA measurement problems in close proximity of the wall, which is also mentioned by Turnow [146]. The mean velocity profiles indicate that the strongest velocity gradients arise at the level of lower channel wall. Additionally, the instabilities in this region (i.e., within the shear layer) result in strong vortices and therefore in high velocity fluctuations, which can be observed in the profiles of the streamwise velocity fluctuation, shown in Fig. 7.9.

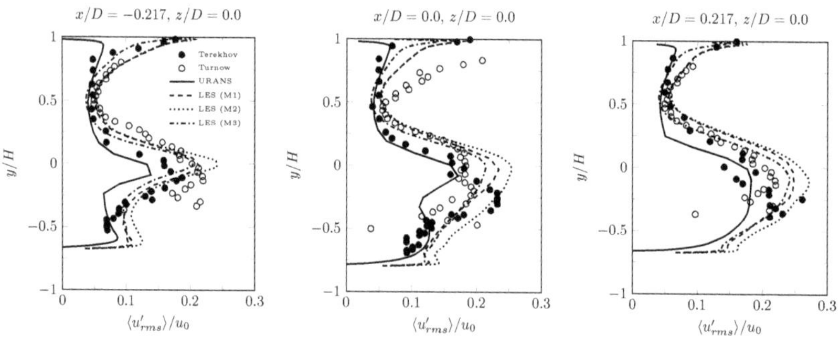

Fig. 7.9: Time-averaged streamwise velocity fluctuation profiles $\langle u'_{rms}\rangle/u_0$ obtained from LES and RANS for different streamwise positions $x/D = 0, \pm0.217$ in the midplane of the channel ($z/D = 0$) in comparison with LDA measurements of Terekhov et al. [143] and Turnow et al. [146].

A deviation between the measured and calculated location of the time-averaged maximum streamwise velocity fluctuations $\langle u'_{rms} \rangle$ is noticeable within the center of the dimple (i.e., at position $x/D = 0$), whereas the profiles at $x/D = \pm 0.217$ are in close agreement to the LDA measurements. Previous URANS calculations in combination with different closure models (e.g., $k - \epsilon$ model and $k - \omega$ SST model), highlighted the main drawback of this turbulence modeling approach, namely the underestimation of velocity fluctuations or turbulence. This becomes clearly visible especially for the near-wall region and in the interface between the channel and the dimple cavity, where the time-averaged streamwise velocity fluctuations were not captured correctly. However, the results presented in Figs. 7.8 and 7.9 confirm that the LES is capable of predicting the flow field for a turbulent channel flow over a single spherical dimple correctly and can be applied for further investigations of the thermo-hydraulic performance and the prediction of particulate fouling.

7.1.3 Thermo-hydraulic performance analysis

For the analysis of the thermo-hydraulic performance of different structured heat transfer surfaces, the pressure loss is determined and expressed through the *Darcy* friction factor[1]:

$$f = -\frac{(\mathrm{d}p \,/\, \mathrm{d}x) D_h}{\rho_f u_b^2 / 2}, \tag{7.1}$$

where $(\mathrm{d}p \,/\, \mathrm{d}x)$ represents the pressure gradient in streamwise direction, D_h is the hydraulic diameter of the channel, ρ_f and u_b are the density and the bulk velocity of the fluid, respectively. The convective heat transfer is evaluated in terms of the local Nusselt number:

$$Nu = \frac{hL}{k} = \frac{q(2H)}{\left(T_w - T_f\right)k}, \tag{7.2}$$

with the convective heat transfer coefficient h, the characteristic length L, the heat flux q and the thermal conductivity k of the fluid. In order to compare the numerical results with empirical correlations, a spatial-averaged Nusselt number $\overline{Nu}$ is calculated as follows:

$$\overline{Nu} = \frac{1}{A} \int_A Nu \, \mathrm{d}A \,, \tag{7.3}$$

where A is the heat transfer surface area. The friction coefficient f_0 and Nusselt number Nu_0 of the corresponding smooth channel are determined using the well-known correlations for pipe flows proposed by Petukhov and Gnielinski [61]:

$$f_0 = (0.790 \ln(Re_d) - 1.64)^{-2}, \tag{7.4}$$

$$Nu_0 = \frac{(f_0/8)(Re_d - 1{,}000)Pr}{1 + 12.7(f_0/8)^{1/2}\left(Pr^{2/3} - 1\right)}. \tag{7.5}$$

[1]The *Darcy* friction factor is directly related to the *Fanning* friction factor by $C_f = f/4$.

These relations were originally formulated in terms of Reynolds number Re_d, based on the tube or pipe diameter d, and are valid for the range of $3{,}000 \lesssim Re_D \lesssim 5 \times 10^6$ and $0.5 \lesssim Pr \lesssim 2{,}000$. Moreover, Eq. (7.5) applies for an uniform surface heat flux as well as for an uniform surface temperature [61]. In order to employ both correlations for channel flows, the Reynolds number Re_d is replaced by the appropriate Reynolds number based on the hydraulic diameter $D_h = 2H$ for a smooth infinitely wide channel [115]. Additionally, the validity of Eqs. (7.4) and (7.5) was proven numerically for different Reynolds numbers (i.e., bulk flow velocities), as shown in Fig. 7.10.

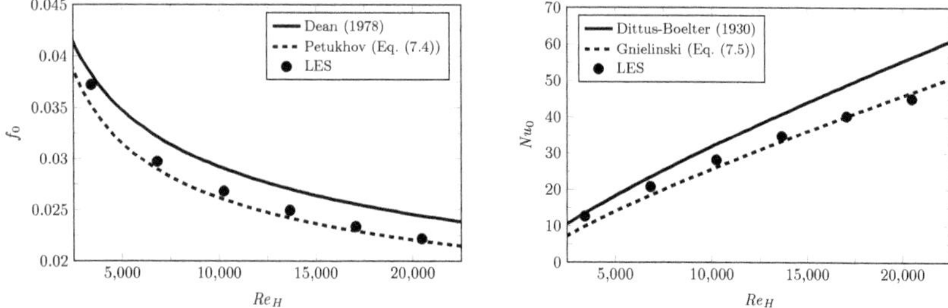

Fig. 7.10: Friction factor f calculated from Eq. (7.4) and Dean's correlation [31] compared to results (left); Nusselt number Nu obtained from Eq. (7.5) and the Dittus-Boelter correlation [61] in comparison to LES results (right).

The time-averaged pressure loss, integral convective heat transfer (integrated over the heat transfer surface area from $x/D = -0.75$ to $x/D = 1.75$ in streamwise direction and from $z/D = -0.75$ to $z/D = 0.75$ in spanwise direction), and thermo-hydraulic efficiency for the clean structured surfaces are summarized and compared in Tab. 7.4 with respect to the smooth channel configuration.

Tab. 7.4: Thermo-hydraulic performance analysis for different clean structured surfaces at $\mathrm{Re}_D = 42{,}000$.

	t/D	f/f_0	$\overline{Nu}/Nu_0$	$\frac{\overline{Nu}/Nu_0}{(f/f_0)}$	$\frac{\overline{Nu}/Nu_0}{(f/f_0)^{1/3}}$
Square cavity	0.26	1.600	1.292	0.808	1.105
Spherical dimple	0.26	1.079	1.508	1.398	1.470
Spherical dimple	0.35	1.338	1.661	1.241	1.507

The evaluation of both types of structured surfaces (i.e., the square cavity and dimple) discloses the disadvantages of the square cavity due to the significant increase of pressure loss f/f_0 by approx. 60%, which cannot be compensated by the slightly enhanced heat transfer $\overline{Nu}/Nu_0$ of around 29%. The thermo-hydraulic efficiency ranges between 0.808

and 1.105, depending on the applied definition[2]. In case of the shallow spherical dimple ($t/D = 0.26$), a greatly enhanced thermo-hydraulic performance can be observed, which is between 1.398 and 1.470. This achievement results from the moderate increase of pressure loss by almost 8%, which agrees with published results by Turnow [146], in comparison to the substantial heat transfer augmentation of about 50%. The pressure loss for the deep dimple ($t/D = 0.35$) is about 24% higher compared to the shallow dimple, whereas the convective heat transfer is ca. 10% higher as for the shallow one, resulting in a thermo-hydraulic performance between 1.241 and 1.507. Thus, the best thermal performance parameter $\overline{Nu}/Nu_0/(f/f_0)$ is achieved by the shallow dimple, whereas a slightly better thermo-hydraulic efficiency $\overline{Nu}/Nu_0/(f/f_0)^{1/3}$ is obtained for the deep dimple. Fig. 7.11 illustrates the time-averaged normalized Nusselt number distribution $\langle Nu \rangle / Nu_0$ for both types of structured surfaces, where regions of high convective heat transfer rates are primarily obtained within the back portion of the surface structures and in the downstream area behind the back rim of the square cavity and dimple.

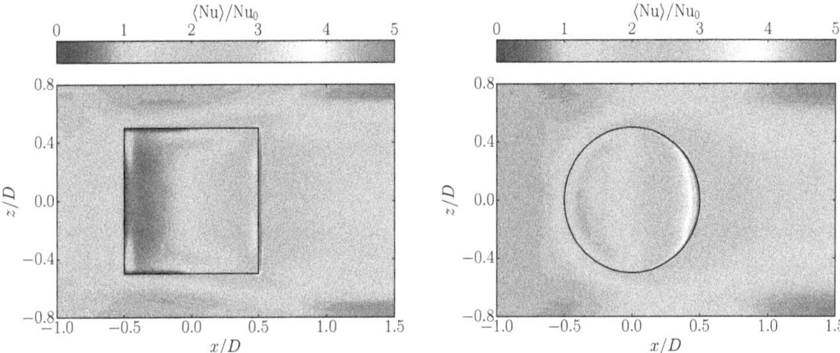

Fig. 7.11: Time-averaged Nusselt number distribution $\langle Nu \rangle / Nu_0$ at the clean lower channel wall for a square cavity (left) and a spherical dimple (right) at $Re_D = 42{,}000$.

Regions of low heat transfer are clearly visible in the front portion of both square cavity and dimple, where low-speed recirculating flows and flow separation occur. An important finding in case of the spherical dimple is that the oscillating vortex structure (see section 7.1.2), which leaves the dimple as ejecting jet at the back rim, does not seem to be the driving factor for the greatly enhanced heat transfer, since the extreme positions of the switching vortex ($\pm 45°$) are not visible in the distribution of the time-averaged Nusselt number (see

[2] The $\overline{Nu}/Nu_0/(f/f_0)$ parameter is referred to as the Reynolds analogy performance parameter. The $\overline{Nu}/Nu_0/(f/f_0)^{1/3}$ parameter is referred to as thermal performance parameter (or thermo-hydraulic efficiency), which is suggested by Gee and Webb [49] to provide a heat transfer augmentation quantity $(\overline{Nu}/Nu_0)$ and a friction factor augmentation quantity $\left((f/f_0)^{1/3}\right)$, where each is given for the same ratio of mass flux in an internal passage with augmentation devices to mass flux in an internal passage with smooth surfaces [83].

Fig. 7.11). Instead, the heat transfer process is mainly augmented by the strong upwash flow at the back rim of the dimple, which greatly enhances the turbulent mixing between the hot fluid in the near-wall region and the coolant mainstream flow in the channel. This effect is illustrated in Fig. 7.12, where the normalized turbulence kinetic energy (TKE) k/u_b^2 along the centerline of the dimpled channel ($z/D = 0$) at $y/D = 0.08$ is compared to the time-averaged heat transfer augmentation $\langle Nu \rangle / Nu_0$ along the dimpled surface.

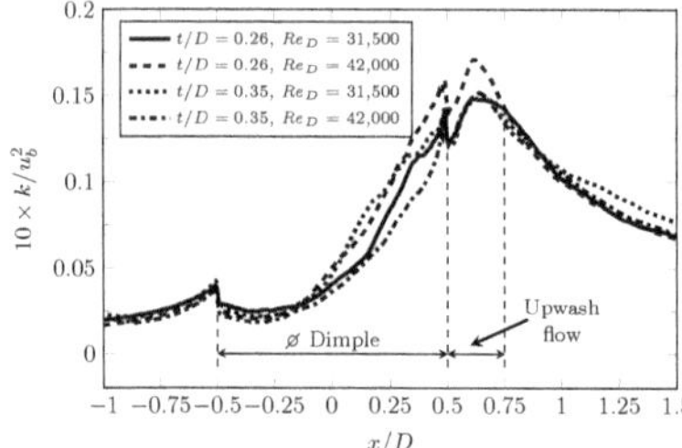

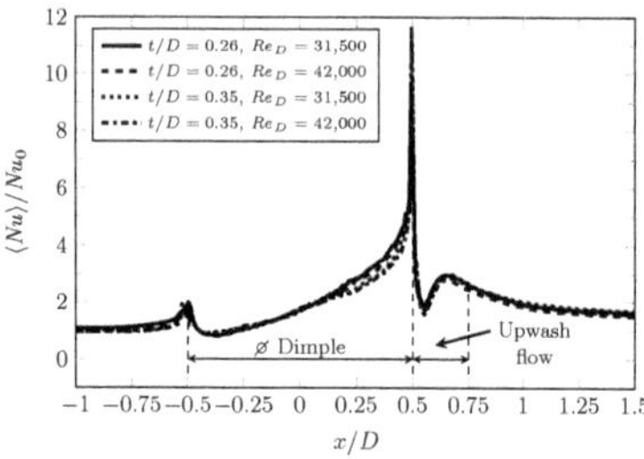

Fig. 7.12: Normalized turbulence kinetic energy k/u_b^2 along the centerline of the dimpled channel ($z/D = 0$) at $y/D = 0.08$ (left) and time-averaged heat transfer augmentation $\langle Nu \rangle / Nu_0$ along the dimpled surface (right).

Regions with high TKE values can be seen clearly near the back rims of the dimples. The strongest upwash flow is generated by the shallow dimple ($t/D = 0.26$) and Reynolds number of $Re_D = 42{,}000$, since the highest TKE values are observed for this case. The slightly increased values of TKE at the front rim of the dimples can be explained by velocity fluctuations induced by flow separation. This leads also to an enhancement of the heat transfer up to $\langle Nu \rangle / Nu_0 = 2.5$, which can be seen in Fig. 7.12, followed by decreased heat transfer rates in the recirculating zone, since the fluid is heated up resulting in an decrease of the temperature difference and therefore a reduction of the heat flux. The high-speed coolant flow impinge in the back portion of the dimple which causes the upwash flow and high local heat transfer rates up to $\langle Nu \rangle / Nu_0 = 11.5$ at the back rim of the dimple. Furthermore, the LES results show a fairly high heat transfer augmentation of about $\langle Nu \rangle / Nu_0 = 1.5$ even at $x/D = 1.5$ (i.e., one dimple diameter behind the trailing edge). For the sake of completeness, it must be mentioned that the switching mono-core vortex structure inside the dimple is not mainly responsible for the superior heat transfer, but contributes to the turbulent mixing at the back rim of the dimple and thus to the heat transfer enhancement, which can be seen in Fig. 7.3, where the location for the highest values of TKE corresponds well with the extreme positions of the oscillating ejecting jet ($\pm 45°$). Moreover, it can be shown that this oscillating jet rotates around its x-axis (i.e., around the streamwise direction) and therefore drives the entrainment of high-speed coolant flow from the core region of the channel into the near-wall region and promoting the ejecting of low-speed hot flow out of the near-wall region, which additionally improves the heat transfer process.

7.1.4 Prediction of particulate fouling

The two different types of heat transfer surfaces, namely the square cavity and the spherical dimple, are further analyzed regarding their susceptibility for particulate fouling, considering different particle mass loadings up to $\eta = 0.2\%$. The simulated fouling time interval is restricted to 120 s of physical real-time due to the high computational costs of multiphase Eulerian-Lagrangian LES. The conducted thermo-hydraulic performance analysis for each case is summarized in Tab. 7.5.

Tab. 7.5: Thermo-hydraulic performance analysis for different fouled structured surfaces and mass loadings η after 120 s of physical real-time at $Re_D = 42{,}000$.

	t/D	η [%]	f/f_0	$\overline{Nu}/Nu_0$	$\frac{\overline{Nu}/Nu_0}{(f/f_0)}$	$\frac{\overline{Nu}/Nu_0}{(f/f_0)^{1/3}}$
Square cavity	0.26	0.1	1.621	1.287	0.794	1.100
Spherical dimple	0.26	0.1	1.075	1.505	1.400	1.469
Spherical dimple	0.35	0.1	1.342	1.658	1.235	1.503
Square cavity	0.26	0.2	1.622	1.290	0.795	1.098
Spherical dimple	0.26	0.2	1.083	1.508	1.392	1.468
Spherical dimple	0.35	0.2	1.343	1.655	1.232	1.501

This analysis is less unambiguous as for the clean surfaces and can be explained by the relatively small change of pressure loss and convective heat transfer, caused by the relatively thin fouling layers after 120 s of physical real-time. Hence, the present results provide a trend concerning the fouling behavior of the investigated structured surfaces. According to this the square cavity is more susceptible for fouling deposits, which leads to an increase of pressure loss of approx. 2% and a reduced heat transfer of about 0.5%. The spherical dimples with a corresponding dimple depth-to-dimple diameter ratio of $t/D = 0.26$ and 0.35 show only minor changes in pressure loss and an almost stable convective heat transfer for all considered particle mass loadings η, indicating a significant lower susceptibility regarding particle depositions or a higher fouling-mitigation potential compared to the square cavity. Fig. 7.13 shows the distribution and thickness x_f of the settled fouling layer for the square cavity and the shallow dimple ($t/D = 0.26$) after 120 s of physical real-time and for a foulant particle mass ratio of $\eta = 0.2\%$. The thermo-physical properties of the randomly injected particles can be found in Tab. 7.3. A relatively high amount of particle deposits with a total fouling layer thickness in the range of about $x_f = 20\,\mu$m to $100\,\mu$m can be observed within the recirculation zone of the spherical dimple and in the vicinity of the front rim or leading edge, whereas no fouling is detected on the back portion of the dimple, where the impingement and reattachment points are located. In case of the square cavity, the highest fouling deposition occurs also inside the forward half of the cavity where the a low-speed recirculating flow promotes the deposition of slow particles. Both surfaces structures generate regions with less or without formations of fouling deposition

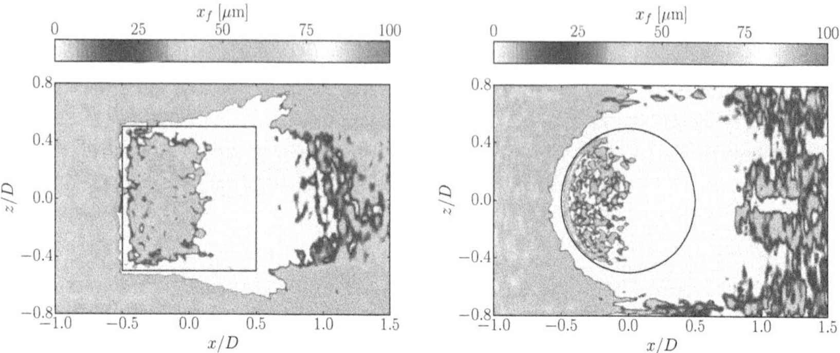

Fig. 7.13: Distribution and height x_f of the fouling layer after 120 s of physical real-time for an particle mass loading $\eta = 0.2\%$ and particle diameter $D_p = 20\,\mu$m at $Re_D = 42{,}000$: square cavity (left), spherical dimple (right).

directly behind the trailing edge or back rim. On the one hand, the reduction of fouling in this zone is closely related to the upwash flow and the resulting turbulent mixing which prevents particles from deposition [134]. On the other hand, the occurring wall shear stress τ_w in streamwise flow direction seems to be the driving factor for the prevention and removal of fouling, especially in case of the spherical dimple shown in Fig. 7.14, where regions of higher wall shear stresses correspond to regions with less or without fouling.

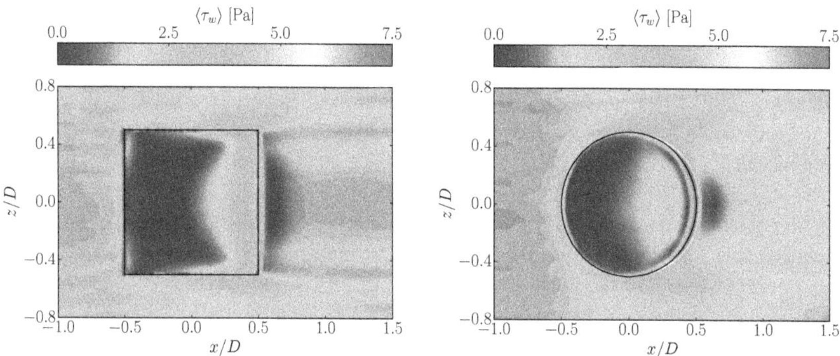

Fig. 7.14: Time-averaged wall shear stress distribution $\langle \tau_w \rangle$ at the clean lower channel wall for a square cavity (left) and a spherical dimple (right) at $Re_D = 42{,}000$.

Considering only the dimpled surfaces, the deep dimple ($t/D = 0.35$) (see Fig. 7.16) exhibits three times higher deposition rates within the low-speed recirculating flow. Additionally, a higher amount fouling deposits is obtained at the channel wall downstream of the dimple trailing edge. This indicates that shallow dimple with a corresponding dimple depth-to-dimple diameter ratio of $t/D = 0.26$ shows a higher fouling-mitigation potential, which will now be further analyzed.

The reason for the increased amount of particulate matters in the front portion of the deep dimple can be easily explained by the local flow conditions in this specific region. Fig. 7.15 shows the time-averaged streamwise velocity profiles within the shallow and deep dimple at different positions ($x/D = 0, \pm 0.217$ and $z/D = 0$) for both the continuous phase (i.e., the carrier flow) and the dispersed phase (i.e., the foulant particles).

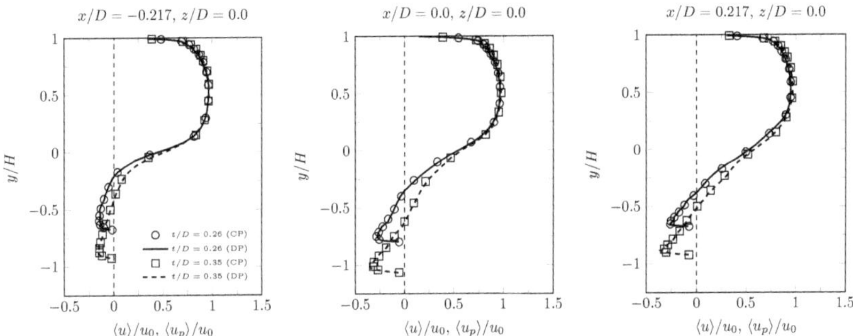

Fig. 7.15: Time-averaged streamwise fluid velocity profiles $\langle u \rangle / u_0$ (continuous phase (CP)) and particle velocity profiles $\langle u_p \rangle / u_0$ (dispersed phase (DP)) obtained from LES for different streamwise positions $x/D = 0, \pm 0.217$ in the midplane of the channel ($z/D = 0$).

It is obvious that the size of the recirculating low-speed flow zone, indicated by the intersection points of the velocity profiles and the dashed line at $\langle u \rangle = \langle u_p \rangle = 0$, is larger for the deep dimple configuration which consequently enlarge the region with an increased probability regarding particle depositions. In addition, the streamwise velocity profiles of both phases are in very close agreement as expected for the injected light foulant particles (i.e., $St \lesssim 1$). Since the streamwise backflow velocity within the deep dimple is higher, foulant particles are caught in the zone of the recirculating low-speed flow and cannot entrain the impingement region in the back portion of the dimple or the mainstream flow within the channel. The increased amount of fouling deposits on the centerline ($z/D = 0$) directly behind the deep dimple is caused by the upwash flow which leaves the dimple at the back rim and is more inclined in case of a dimple depth-to-dimple diameter ratio of $t/D = 0.35$ as for $t/D = 0.26$, leading to the formation of a small region where flow separation occurs (i.e., low streamwise flow velocities (see Fig. A.5)) which promotes the deposition of particles in the downstream area directly behind the deep dimple, as shown in Fig. 7.16. Moreover, Fig. 7.16 shows a comparison between the fouling layer distribution obtained by LES and URANS, revealing the most probable reason for the high fouling mitigation or self-cleaning effect induced by dimples mentioned by Kasper et al. [67, 66, 68] and Deponte et al. [35, 34]. Since URANS is not suitable for the simulation of vortex switching inside the dimple and therefore the oscillating outflow (see Fig. 7.5) at the back

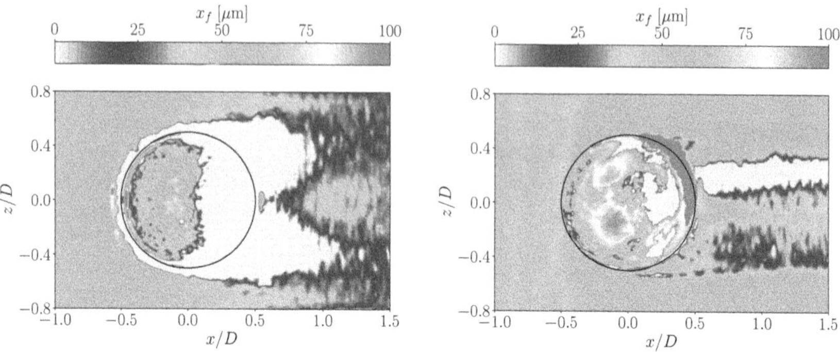

Fig. 7.16: Distribution and height x_f of the fouling layer for the deep dimple ($t/D = 0.35$) after 120 s of physical real-time for a particle mass loading $\eta = 0.2\%$ and particle diameter $D_p = 20\,\mu\text{m}$ at $Re_D = 42{,}000$: LES (left), URANS (right).

rim of the dimple, the simulated fouling layer from URANS differs completely from that calculated by LES. Thus, the oscillating ejecting jet can be identified as the main driver for the prevention and removal of particulate fouling. This is an observation which corresponds well with Figs. 7.13, 7.14, and 7.16, where the extreme positions of the alternating or oscillating vortex structure ($\pm 45°$) become visible in both, the fouling and wall shear stress distribution. Fig. 7.17 presents the time-averaged streamwise velocity $\langle u \rangle / u_b$ in the $y - z$ plane at $x/D = 0.5$ (i.e., at the back rim) for both dimple configurations.

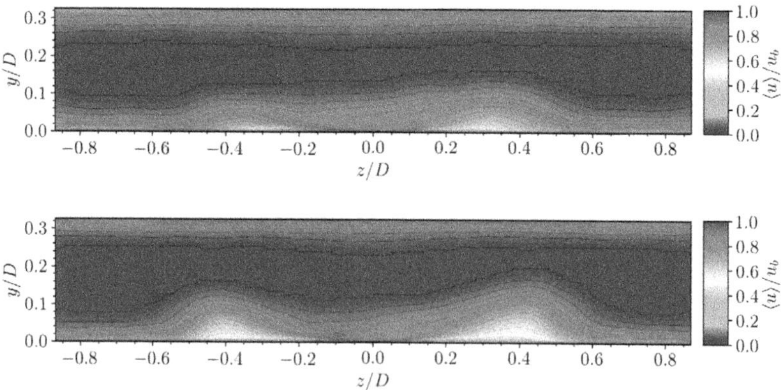

Fig. 7.17: Distribution of the time-averaged streamwise velocity $\langle u \rangle / u_b$ in the $y - z$ plane at $x/D = 0.5$ for $Re_D = 42{,}000$ and a dimple depth-to-dimple diameter ratio of $t/D = 0.26$ (top) and $t/D = 0.35$ (bottom).

It shows clearly the extreme position of the oscillating outflow for the spanwise direction at $z/D \approx \pm 0.35$ in case of the shallow dimple and at $z/D \approx \pm 0.4$ for the deep dimple, which again agrees well with the observed fouling layers. However, the time-averaged velocity distribution reveals that switching outflow leaves the dimple at the back rim with a significantly lower streamwise velocity compared to the undisturbed high-speed mainstream flow in the channel, since the mono-core vortex structure looses its momentum within the dimple cavity. Thus, the local increase of the wall shear stresses cannot be explained by a local increase of the streamwise flow velocities, but rather by the rotation and location (i.e., distance to the lower wall) of these alternating vortex structures, as shown in Fig. 7.18.

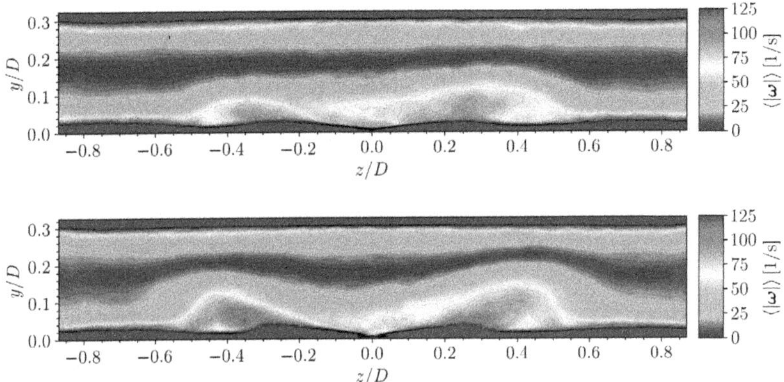

Fig. 7.18: Distribution of the time-averaged vorticity magnitude $\langle|\boldsymbol{\omega}|\rangle$ in the $y-z$ plane at $x/D = 0.5$ for $Re_D = 42{,}000$ and a dimple depth-to-dimple diameter ratio of $t/D = 0.26$ (top) and $t/D = 0.35$ (bottom).

The time-averaged distribution of the vorticity magnitude $\langle|\boldsymbol{\omega}|\rangle$ in the $y-z$ plane at $x/D = 0.5$ (i.e., directly behind the dimple) shows a high amount of vorticity in the upper and lower near-wall region generated by the turbulent boundary layer, but also regions with a greatly enhanced vorticity at the corresponding extreme positions of the oscillating outflow (i.e., at $z/D \approx \pm 0.35$ and $z/D \approx \pm 0.4$), indicating a strong rotation of these mono-core flow structures. Additionally, the wall-normal distance (y) between the lower channel wall and the maximum values of the time-averaged vorticity magnitude $\langle|\boldsymbol{\omega}|\rangle$ is larger for the deep dimple ($t/D = 0.35$). It was originally shown by Kravchenko et al. [76], by means of DNS for a fully developed turbulent channel flow at $Re_\tau = 180$, that high-skin friction regions can be directly attributed to streamwise vortices. Therefore, they introduced the two-point correlation of the wall shear rate $\partial u/\partial y$ and the streamwise vorticity ω_x as follows:

$$R(r_x, y, r_z) = \left\langle \frac{\partial u}{\partial y}(x, 0, z)\omega_x(x + r, y, z + r_z) \right\rangle, \tag{7.6}$$

where r_x and r_z are the spatial separations in x- and z-directions (see also section 2.2), and $\langle ... \rangle$ denotes an average over x, z and time. It was pointed out by Kravchenko et al. [76] that skin-friction on the wall is mostly correlated with downstream streamwise vorticities. Moreover, the location of the local maximum of the (conditionally averaged) streamwise vorticity is closer to the wall than the maximum location of the two-point correlation, Eq. (7.6), indicating that streamwise vortices associated with higher wall shear rates are closer to the wall. The latter observation is of major importance, since it provides an explanation for both, the increase of the local wall shear stresses due to the switching vortex structures and the superior fouling performance of the shallow dimple. Fig. 7.19 presents the time-averaged streamwise vorticity $\langle \omega_x \rangle$ in the $y-z$ plane at $x/D = 0.5$.

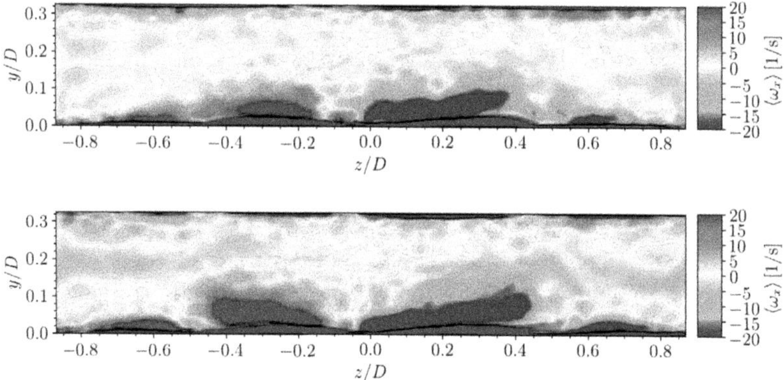

Fig. 7.19: Distribution of the time-averaged streamwise vorticity $\langle \omega_x \rangle$ in the $y-z$ plane at $x/D = 0.5$ for $Re_D = 42{,}000$ and a dimple depth-to-dimple diameter ratio of $t/D = 0.26$ (top) and $t/D = 0.35$ (bottom). Positive values of $\langle \omega_x \rangle$ indicates counter-clockwise rotation whereas negative values of $\langle \omega_x \rangle$ indicates clockwise rotation.

The distributions of the time-averaged streamwise vorticity $\langle \omega_x \rangle$ disclose the existence of different high-vorticity regions, whereby positive values of streamwise vorticity indicate, according to the right-hand rule, counter-rotating rotation of the fluid flow whereas negative values indicates clockwise rotation. The enhanced streamwise vorticity in the near-wall region is a consequence of the no-slip boundary condition at the channel wall and thus of the presence of the turbulent boundary layer. However, two additional areas of high vorticity are visible above the near-wall region at the lower channel wall, where the location of these particular regions clearly corresponds with the location of the switching mono-core vortex structure which leaves the dimple cavity at the trailing edge. While the absolute values of the streamwise vorticity are almost similar for both dimple configurations, the distance between the lower channel wall and the high-vorticity regions differs. For the shallow dimple ($t/D = 0.26$), the location of enhanced streamwise vorticity is located closer to the wall, compared to the deep dimple ($t/D = 0.35$), where the steep wall in the back portion of the dimple inclines the outflow and therefore shifts the maximum of the streamwise vorticity away from the lower channel wall. According to Kravchenko et al. [76], this results in higher local wall shear stresses in the downstream area of the shallow dimple, and therefore a stronger self-cleaning process and reduced particulate fouling, which is in good agreement to the predicted fouling layer distributions, shown in Figs. 7.13 and 7.16. The potential to mitigate particulate fouling is further analyzed for both dimple configurations in the following section 7.1.5, including the experimental confirmation of the numerical results.

7.1.5 Experimental confirmation of numerical results

In order to obtain further insights into the fouling behavior of both dimple configurations and to confirm the numerical results, the predicted particulate fouling is compared to experimental results from laboratory scale fouling test conducted by ICTV at the Technische Universität Braunschweig. Detailed information regarding the fouling test rig and experimental setup can be found in Deponte et al. [35, 34] and Kasper et al. [66]. Since, it is not possible to calculate the thermal fouling resistance $R_{f,th}$ from the conducted fouling simulations[3], the mass-based fouling resistance $R_{f,m}$ is introduced for the sake of comparison. Accordingly, the specific mass of the fouling layer m_f and the density of the foulant particles ρ_p are used to calculate the mean height x_f of the fouling layer. This value is corrected by the packing factor $(1-\epsilon)$, which is determined from void fraction ϵ of the fouling layer. Due to the high sphericity of the particles used in the conducted experiments as well as the perfectly spherical-shaped particles injected during the simulations, a sphere packing within the fouling layer can be assumed. For a random packing, the packing factor $(1-\epsilon)$ becomes 0.64 [131]. Hence, the height of the fouling layer can be defined as:

$$x_f = \frac{m_f}{\rho_p(1-\epsilon)}, \tag{7.7}$$

The local height of the fouling layer x_f and the corresponding thermal conductivity $k_{f,tot}$ are used to estimate the mass-based fouling resistance [15]:

$$R_{f,m} = \frac{x_f}{k_{f,tot}}. \tag{7.8}$$

Since the fouling layer consists of both, deposited foulant particles and water-filled cavities, the overall thermal conductivity $k_{f,tot}$ is considered as a series-connected resistance[4]:

$$\frac{1}{k_{f,tot}} = \frac{1-\epsilon}{k_p} + \frac{\epsilon}{k_w}, \tag{7.9}$$

[3]The heat conduction through the lower and upper channel wall is not simulated by conjugate heat transfer (CHT). Instead, a uniform-temperature BC was directly set at the walls (see Tab. 7.1).

[4]The assumption of a series-connected resistance is valid for a random spherical packing of homogeneous powdered materials with a point contact between the particles [77].

where k_p and k_w are the thermal conductivities of the foulant particles and water, respectively. Substitution of Eqs. (7.7) and (7.9) into Eq. (7.8) yields the mass-based fouling resistance:

$$R_{f,m} = \frac{m_f}{\rho_p(1-\epsilon)}\left(\frac{1-\epsilon}{k_p} + \frac{\epsilon}{k_w}\right). \tag{7.10}$$

Besides the thermal and mass-based fouling resistance $R_{f,th}$ and $R_{f,m}$, respectively, an alternative method for the quantification of particulate fouling is suggested by Deponte et al. [33], namely the *Phosphorescent Fouling Quantification* (PFQ). This method provides access to the local distribution of the fouling resistance by optically determining the height of the fouling layer using phosphorescent particles (Lumilux green SN-F5).
For an optical or qualitative confirmation of the numerical results through experimental results, a comparison between the simulated and measured fouling layer distribution is conducted for a fully turbulent channel flow over the shallow spherical dimple with a dimple depth-to-dimple diameter ratio of $t/D = 0.26$. The thermo-physical properties of the dispersed and continuous phase in the numerical simulations are adapted to match the properties from the experimental investigations. In contrast to the experimental setup, periodic boundary conditions in spanwise direction are applied, since experimental investigations have shown a negligible influence of the channel side walls, and therefore of the resulting turbulent boundary layer, on the fouling layer in the channel core flow for the chosen channel geometry (i.e., channel width). The comparison of the numerical and experimental results is given in the overlay Fig. 7.20, where the direction of fluid flow is from the upper left corner to the bottom right one. All three characteristic regions of the experimental fouling layer distribution, as described in Kasper et al. [66], can be reproduced by the simulations with a sufficient accuracy. The relatively high fouling layer thickness in the undisturbed channel flow area (1) upstream of the dimple front rim, the particle deposition within the low-speed recirculation zone (2) inside of the dimple, and the region with no or even less particle deposition downstream of the dimple back rim or trailing edge (3) become clearly visible and are captured well with a good agreement.

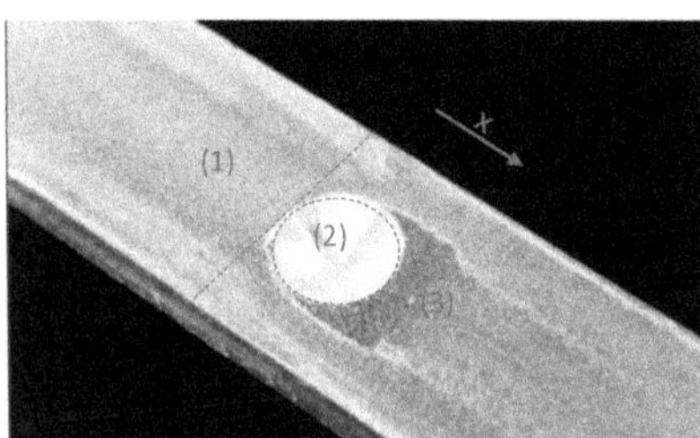

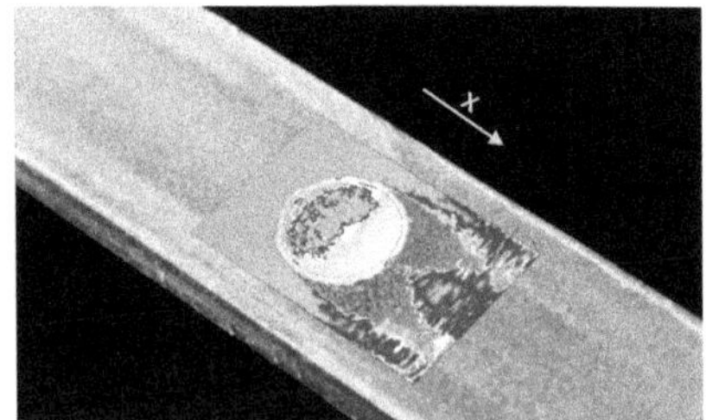

Fig. 7.20: Qualitative comparison between the simulated and experimental fouling layer distribution for particle mass loading $\eta = 0.2\%$ and particle diameter $D_p = 3\,\mu\text{m}$ at $\text{Re}_D = 10{,}000$ (fluid flow is in x-direction). The colored areas are fouling layers with the thickness increasing from blue over green to yellow.

The high fouling-mitigation potential or self-cleaning effect induced by the investigated dimpled surfaces, which is disclosed by the numerical fouling predictions as well as by the experimental fouling layer, is further analyzed quantitatively in Fig. 7.21 in terms of the mass-based fouling resistance $R_{f,m}$, calculated according to Eq. (7.9). In case of the

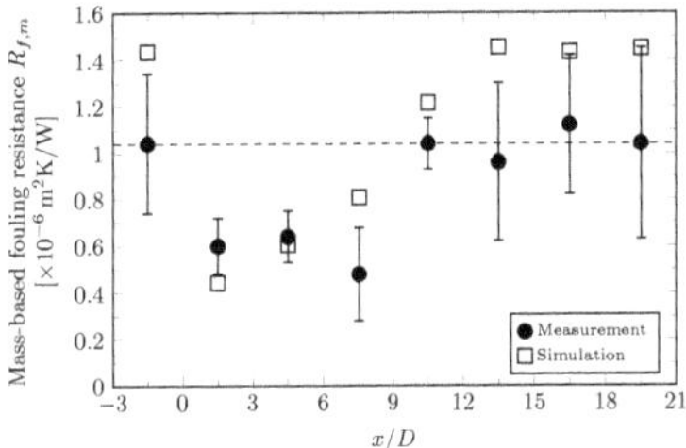

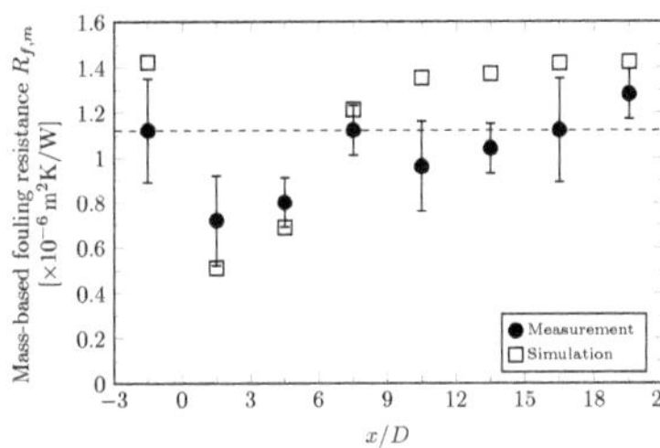

Fig. 7.21: Comparison between the measured and simulated mass-based fouling resistance $R_{f,m}$ around a single dimple with a dimple depth-to-dimple diameter ratio of $t/D = 0.26$ (left) and $t/D = 0.35$ (right) for $Re_D = 10.000$ and $\eta = 0.2\%$. The center of the dimple is located at $x/D = 0$.

shallow dimple ($t/D = 0.26$), the deposition of particles in the undisturbed channel flow at $x/D = -1.5$ (i.e., in front of the dimple) results in a mass-based fouling resistance of about $R_{f,m} = 1.05 \cdot 10^{-6}\,\text{m}^2\text{K/W}$ and is set as reference value (see dashed line in Fig. 7.21). At the following three positions (i.e., $x/D = 1.5$, 4.5 and 7.5), a remarkable reduction of the mass-based fouling resistance is obtained in the experiment [34] and the numerical simulation which is, referred to the reference area, about 43.3% at $x/D = 1.5$ based on the conducted measurements and compared to the reference value of $R_{f,m} = 1.05 \cdot 10^{-6}\,\text{m}^2\text{K/W}$. The region of reduced particulate fouling occurs in streamwise direction from $x/D = 0$ to approx. $x/D = 9$, whereas for $x/D \geq 10.5$, the mass-based fouling resistance $R_{f,m}$ is on the reference level. The overall agreement between measurements and simulation is good and the progression as well as the absolute values of the mass-based fouling resistance, which are within the range of the measurement uncertainty for almost every sample point, are captured well by the simulation. A similar reliable accuracy of the simulation is obtained for the deep dimple ($t/D = 0.35$). Moreover, the measured and simulated progression of $R_{f,m}$ reveal a notable lower potential of fouling mitigation compared to the shallow dimple. Especially the length of the self-cleaning area behind the dimple is reduced and ranges from $x/D = 0$ to approx. $x/D = 7.5$. Based on the measurements, the calculated reduction of fouling is 35.7% at $x/D = 1.5$ which highlights the significantly lower self-cleaning effect induced by the deep dimple. Additionally, Fig. 7.22 shows the time-averaged distribution of the turbulence kinetic energy k/u_b^2 in the $x - z$ plane at $y/D = 0.08$ for the shallow dimple ($t/D = 0.26$), compared to the experimentally generated fouling layer. Clearly, the area of reduced fouling behind the dimple exactly corresponds to the region of enhanced mixing or increased turbulence kinetic energy. This emphasizes the importance of the upwash flow regarding the mitigation of particulate fouling, since high velocity fluctuations and strong turbulent mixing prevents foulant particles from deposition. For completeness,

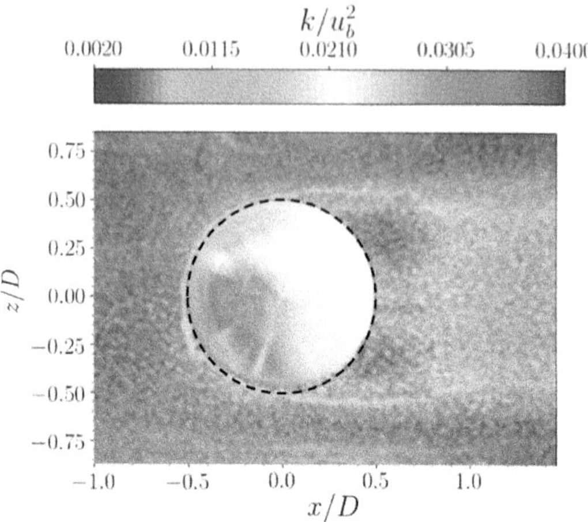

Fig. 7.22: Time-averaged distribution of the turbulence kinetic energy k/u_b^2 in the $x-z$ plane at $y/D = 0.08$ for the shallow $(t/D = 0.26)$ compared to the experimentally generated fouling layer.

Tab. 7.6 summarizes the measured and simulated thermo-hydraulic performance values $Nu/Nu_0/(f/f_0)$ and $Nu/Nu_0/(f/f_0)^{1/3}$ at different Reynolds numbers Re_D for thirteen clean shallow dimples $(t/D = 0.26)$ arranged in a row and a temperature difference of $\Delta T = 20\,\mathrm{K}$.

Tab. 7.6: Measured and simulated thermo-hydraulic efficiency for a row of thirteen clean dimples with $t/D = 0.26$ at $\Delta T = 20\,\mathrm{K}$.

	Experiments (ICTV)				Simulation			
Re_D	Nu/Nu_0	f/f_0	$\frac{Nu/Nu_0}{(f/f_0)}$	$\frac{Nu/Nu_0}{(f/f_0)^{1/3}}$	$\overline{Nu}/Nu_0$	f/f_0	$\frac{\overline{Nu}/Nu_0}{(f/f_0)}$	$\frac{\overline{Nu}/Nu_0}{(f/f_0)^{1/3}}$
10.000	1.328	1.049	1.266	1.307	1.315	1.040	1.264	1.298
20.000	1.420	1.122	1.265	1.367	1.396	1.173	1.190	1.324
30.000	1.448	1.088	1.331	1.408	1.549	1.203	1.287	1.457
40.000	1.461	1.107	1.320	1.412	1.521	1.202	1.265	1.431

The measured and simulated thermo-hydraulic performance values are in close agreement for each investigated Reynolds number Re_D, which underlines the high accuracy of the conducted simulations. Moreover, since the thermo-hydraulic efficiency of a dimple row, consisting of thirteen dimples, can be interpreted as an ensemble-averaged values over thirteen repetitions, the thermo-hydraulic performance values $Nu/Nu_0/(f/f_0)$ and

$Nu/Nu_0/(f/f_0)^{1/3}$ for the dimple row, Tab. 7.22, are directly comparable to those for a clean single spherical dimple at equal Reynolds numbers Re_D (Reynolds analogy). The comparison of the measured $Nu/Nu_0/(f/f_0)^{1/3}$ value for the row of shallow dimples ($t/D = 0.26$) at $Re_D = 40{,}000$ with the $Nu/Nu_0/(f/f_0)^{1/3}$ value for a single shallow dimple at $Re_D = 42{,}000$, given in Tab. 7.4, reveals only a minor deviation of about 2.7% which again highlights the accuracy of the performed simulations.

7.1.6 Summary

Different types of structured heat transfer surfaces, namely the square cavity and a spherical dimple with a corresponding dimple depth-to-dimple diameter ratio of $t/D = 0.26$ (shallow) and $t/D = 0.35$ (deep) have been numerically investigated regarding their pure thermo-hydraulic performance values $Nu/Nu_0/(f/f_0)$ and $Nu/Nu_0/(f/f_0)^{1/3}$ and their susceptibility for particulate fouling. Additionally, the comparison of LDA measurements (Terekhov [143], Turnow [146]) with selected numerical results obtained for the shallow dimple has shown that the preformed LES are able to predict the specific flow phenomena induced by dimpled surfaces (e.g., low-speed recirculating flow, upwash flow, vortex switching) with high accuracy. Considering the clean structured surfaces (i.e., without particulate fouling), the dimpled surfaces show the best thermo-hydraulic efficiency. Depending on the employed definition of the thermal-hydraulic performance, the shallow dimple shows a superior thermo-hydraulic efficiency of $Nu/Nu_0/(f/f_0) = 1.398$ compared to 1.241 obtained for deep dimple. By using the $Nu/Nu_0/(f/f_0)^{1/3}$ value, both dimple configurations show an almost equal thermal-hydraulic performance. The impingement of high-speed mainstream flow in the back portion of the dimple greatly enhances the turbulent mixing between the coolant flow and the hot flow in the near-wall region, which is the main reason for the superior heat transfer enhancement.
For the fouled surfaces, the switching mono-core vortex inside the dimples could be identified as main driver for a self-cleaning process induced through dimpled surfaces, since the streamwise vorticity generated by the oscillating vortex structure, which leaves the dimple cavity at the back rim, leads to an increase of the local wall shear stresses. While the deterioration of the thermo-hydraulic performance is almost equal for both dimple configurations, the predicted fouling layer distributions reveal a higher fouling-mitigation potential in case of the shallow dimple, which is also confirmed by experimental fouling investigations. The comparison between the experimentally generated fouling layer and the numerically predicted fouling layer reveals similar characteristic regions (e.g., reduced particulate fouling in the downstream area of the dimple). Moreover, the measured and simulated mass-based fouling resistance $R_{f,m}$, which also indicates a better fouling performance of the shallow dimple, is in good agreement and reveal the reliable accuracy of the performed multiphase Eulerian-Lagrangian LES. The main reason for superior fouling-mitigation potential of the shallow dimple can be explained by the location of the high-vorticity regions, which are located closer to the lower channel wall, compared to the deep dimple configuration. This significantly enhances the local wall shear stresses and therefore the self-cleaning effect induces by the shallow dimple.

7.2 Particulate fouling on spherical dimples in a staggered arrangement

The thermo-hydraulic efficiency analysis as well as the prediction of particulate fouling for spherical dimples in a staggered arrangement (dimple package) is presented in this section. Such arrangements of spherical dimples are commonly used in many industrial applications (e.g., turbine cooling, heat exchangers). Thus, this study will expand the fouling investigations to an application with a higher industrial relevance. Two different dimple depth-to-dimple diameter ratios of $t/D = 0.26$ and 0.35 at $Re_D = 42{,}000$ are considered.

7.2.1 Case description and numerical setup

The computational domain for the particle-laden turbulent channel flow over dimples in a staggered arrangement (dimple package) is given in Fig. 7.23, with a size of $9.936H \times H \times 9.936H$ in x-, y- and z-direction and a channel height of $H = 0.015\,\text{m}$. The sharp-edged spherical dimples have a uniform dimple diameter of $D = 0.046\,\text{m}$ and the distance between neighboring dimples is $S = 0.81D$. Two different dimple depth-to-diameter ratios, namely $t/D = 0.26$ and 0.35, were considered in this study.

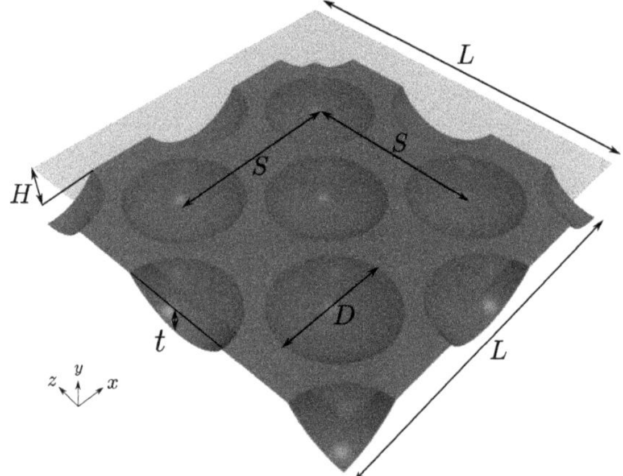

Fig. 7.23: Sketch of the computational domain for the turbulent particle-laden channel flow over sharp-edged spherical dimples in a staggered arrangement.

Periodic boundary conditions are imposed in streamwise (x) and spanwise direction (z), whereas no-slip conditions are applied at the lower and upper channel walls. The structured, bottom wall is heated through a constant, non-dimensional temperature $T^+ = (T - T_\infty) / (T_w - T_\infty) = 1$ and the smooth, opposite channel wall is cooled with

$T^+ = 0$. The molecular Prandtl number Pr is set to 0.71, whereas the turbulent Prandtl number Pr_t is 0.9. The corresponding Reynolds number is $Re_D = u_b D/\nu = 42{,}000$ based on the averaged bulk velocity $u_b = 0.91\,\text{m/s}$ and the dimple print diameter D. The employed boundary conditions are recapped in Tab. 7.7. Based on the LES results,

Tab. 7.7: Boundary conditions applied for the LES of the turbulent channel flow over spherical dimples in a staggered arrangement (dimple package).

Flow variable	Inlet	Outlet	Lower wall	Upper wall	Sides
Velocity	periodic	periodic	$\overline{\mathbf{u}} = 0$	$\overline{\mathbf{u}} = 0$	periodic
Pressure	periodic	periodic	$\partial\overline{p}/\partial\mathbf{n} = 0$	$\partial\overline{p}/\partial\mathbf{n} = 0$	periodic
Temperature	periodic	periodic	$\overline{T^+} = 1$	$\overline{T^+} = 0$	periodic

shown in Fig. 7.24, from a series of preliminary simulations for a smooth channel flow with a corresponding shear Reynolds number of $Re_\tau = 395$ and equal dimensions as for the channel flow with the dimple arrangement at the bottom wall, the computational domain is spatially discretized by a block-structured curvilinear grid consisting of 8,529,920 hexahedral cells. A moderate, linear grid stretching in wall normal direction (y) is used to place the first grid node inside of the laminar sublayer at $y^+ \approx 1$. The modeling of the unresolved subgrid scales is realized through the DOE model. Spherical, mono-dispersed glass beads with a particle diameter of $D_p = 20\,\mu\text{m}$, typical for fouling related problems, and a density ratio of $\rho_p/\rho_f \approx 2.5$ are randomly injected at the domain inlet. Further particle parameters including the Stokes number $St = \tau_p/\tau_f$, calculated by the particle relaxation time $\tau_p = \rho_p D_p^2/18\mu$ and the characteristic time scale of the flow $\tau_f = \nu/u_\tau^2$, are summarized in Tab. 7.8.

Tab. 7.8: Additional particle parameters used in the Euler-Lagrangian LES of the particle-laden turbulent channel flow over dimples in a staggered arrangement, based on the spatial and temporal averaged shear velocity $\langle u_\tau \rangle$ for the (clean) dimpled channel wall.

t/D	u_τ [m/s]	τ_p[s]	τ_f[s]	St
0.26	0.062	5.556×10^{-5}	2.601×10^{-4}	0.2
0.35	0.074	5.556×10^{-5}	1.826×10^{-4}	0.3

Similar to the fouling simulations performed for a single spherical dimple, a particle mass loading of $\eta = \dot{m}_p/\dot{m}_f = 0.2\%$ is chosen to ensure an asymptotic fouling layer growth and steady-state fouling conditions within a few minutes of physical real time. Particle motions are computed considering drag, lift, gravity and buoyancy, added mass as well as thermophoresis and pressure gradient force. Due to the low volume fraction of the dispersed phase, the neglect of inter-particle collisions is assumed and only two-way phase coupling is applied. To ensure a proper domain size large enough to capture all flow and particle structures (see Sardina et al. [125]) the streamwise $\langle C'(x)C'(x+\Delta x)\rangle(y)$ and spanwise two-point correlations $\langle C'(z)C'(z+\Delta z)\rangle(y)$ for the particles at $y/H = 0.15$ (closely above the dimples) are shown in Fig. 7.25. The Auto-correlation function $\rho(x^+)$ and $\rho(z^+)$ indicates only a weak correlation of the particles in streamwise and spanwise direction. Additionally, since $\rho(x^+)$ and $\rho(z^+)$ decay to zero, the chosen domain size is proved to be large enough to capture the occurring particle structures.

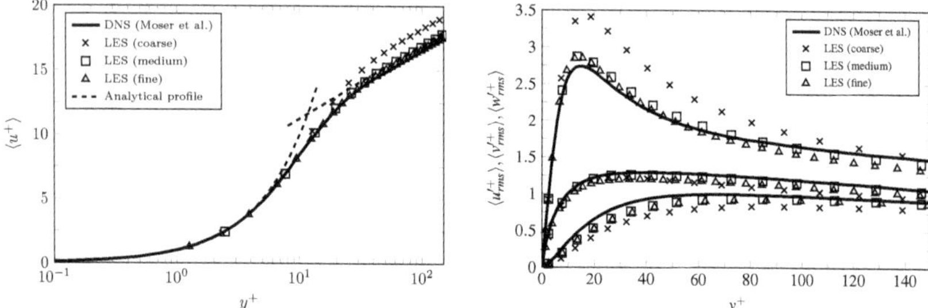

Fig. 7.24: Mean streamwise fluid velocity profile $\langle u^+ \rangle$ and rms fluid velocity profiles $\langle u'^+_{rms} \rangle$, $\langle v'^+_{rms} \rangle$ and $\langle w'^+_{rms} \rangle$ in wall units compared to the DNS data provided by Moser et al. [107].

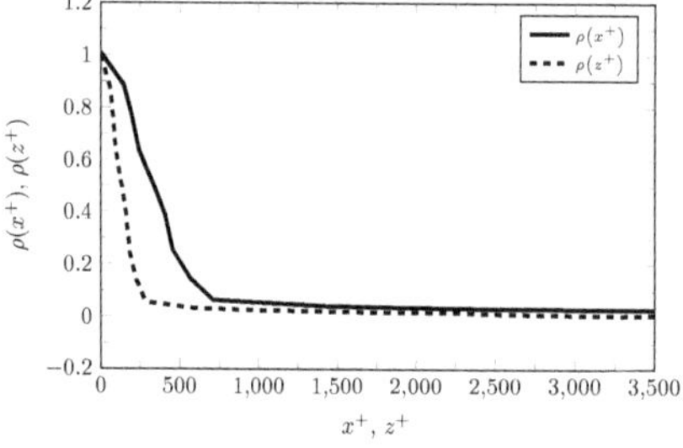

Fig. 7.25: Spatial streamwise and spanwise auto-correlation functions $\rho(x^+)$ and $\rho(z^+)$ of the particle concentration fluctuations C' obtained for dimple package $t/D = 0.35$, along the x- and z-domain midplane at $y/H = 0.15$.

7.2.2 Thermo-hydraulic performance analysis

An analysis of the thermo-hydraulic performance, including the investigation of the interaction between particulate fouling and local flow structures, has been performed for sharp-edged spherical dimples in a staggered arrangement using Eulerian-Lagrangian LES with a corresponding mass loading of $\eta = 0.2\%$. Tab. 7.9 presents the thermo-hydraulic performance analysis based on the obtained LES results and the empirical (smooth channel) correlations, Eqs. (7.4) and (7.5), for the investigated clean dimple packages. Significant influence of the dimple depth becomes very clear, since the additional hydraulic loss f/f_0 due to the deep dimples ($t/D = 0.35$) is two times higher, compared to the shallow dimples ($t/D = 0.26$). Contrary to this observation, the influence of the dimple depth on the heat transfer enhancement $\overline{Nu}/Nu_0$ is notably smaller, whereby the heat transfer augmentation in case of the deep dimples is about 20% greater as for the shallow dimples, resulting in a slightly better thermo-hydraulic performance (approx. 5%) of the shallow dimples. Accordingly, the increase of the additional hydraulic loss caused by the larger dimple depth cannot be compensated by the heat transfer augmentation. This observations are generally in line with results published by Ligrani et al. [84, 83] and Turnow et al. [149].

Tab. 7.9: Thermo-hydraulic performance analysis for clean dimples in a staggered arrangement at $\mathrm{Re}_D = 42{,}000$.

t/D	f/f_0	$\overline{Nu}/Nu_0$	$\frac{\overline{Nu}/Nu_0}{(f/f_0)}$	$\frac{\overline{Nu}/Nu_0}{(f/f_0)^{1/3}}$
0.26	5.466	3.023	0.553	1.716
0.35	11.074	3.644	0.329	1.635

In Fig. 7.26, the local flow structures caused by the different dimple packages are visualized by streamline patterns phase-averaged over a corresponding time interval of $t = 0.5$ s. This

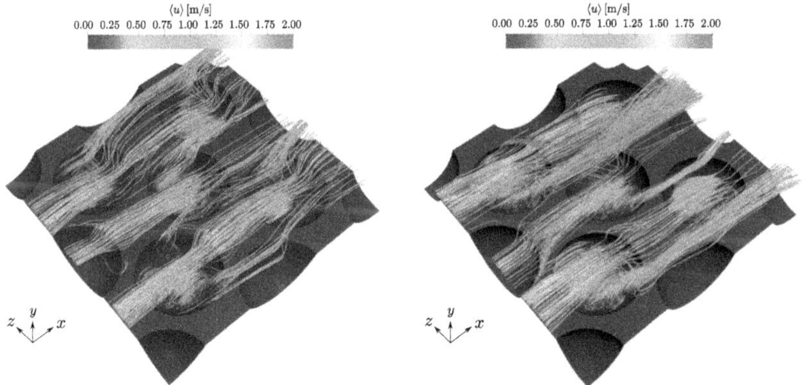

Fig. 7.26: Phase-averaged streamline patterns for $t = 0.5$ s inside a dimple package with different dimple depths at $Re_D = 42{,}000$: $t/D = 0.26$ (left) and $t/D = 0.35$ (right).

illustration shows clearly the strong mixing and high complexity of the flow, which is more or less stochastic without distinctive coherent structures, as reported for the single spherical dimple [67, 66, 147]. The perturbations, coming from the upstream located dimples, destroy and break down the large, organized structures into small scale eddies [149]. Especially for the shallow dimples at $t/D = 0.26$, the identification of coherent structures is very difficult. However, some typical flow structures can be recognized from the phase-averaged streamlines, for instance the reverse flow with a strong recirculation zone inside the dimples. A certain part of the incoming fluid attaches the downstream side and leaves the dimple directly at the trailing edge, while the main part of the arriving flow enters the dimple, rotates and is finally ejected over the dimple side edge into the high-speed mainstream flow. Moreover, the flow structures produced by the deep dimples are more concentrated and causes higher flow velocities, primarily in the channel flow above the dimples. Fig. 7.27 and 7.28 shows the time-averaged convective heat transfer in terms of the Nusselt number $\langle Nu \rangle$ (normalized by the Nusselt number Nu_0 for the smooth, narrow channel) as well as the wall shear stress $\langle \tau_w \rangle$ at the clean, dimpled bottom wall for both dimple package configurations. The highest heat transfer rates are observed,

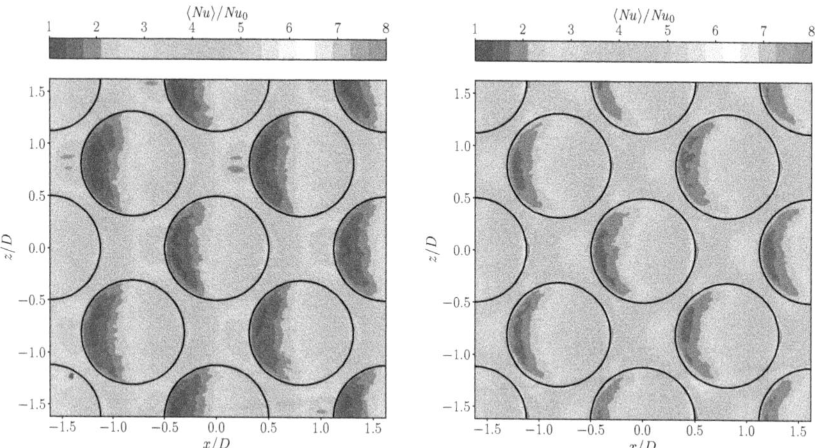

Fig. 7.27: Time-averaged Nusselt number distribution $\langle Nu \rangle / Nu_0$ at the clean, dimpled bottom wall at $Re_D = 42{,}000$: $t/D = 0.26$ (left) and $t/D = 0.35$ (right).

independently from the investigated dimple depth, at the downstream side and trailing edge of the dimple, where the cold flow attaches the surface and maximum flow velocities occur. For the deep dimples it can be up to eight times higher than for the smooth channel. Additionally, the stable zones of low-speed recirculating flows within the dimples are responsible for a remarkable reduction of the heat transfer augmentation within the front portion of the dimples. This region of low heat transfer rate is more pronounced for the shallow dimple package ($t/D = 0.26$), resulting in 20% lower overall heat transfer enhancement in comparison to the deep dimples.In contrast to previous experimental [84] and numerical investigations [149] of dimpled surfaces, no optimal dimple depth-to-diameter ratio concerning a maximum thermo-hydraulic performance could be determined.

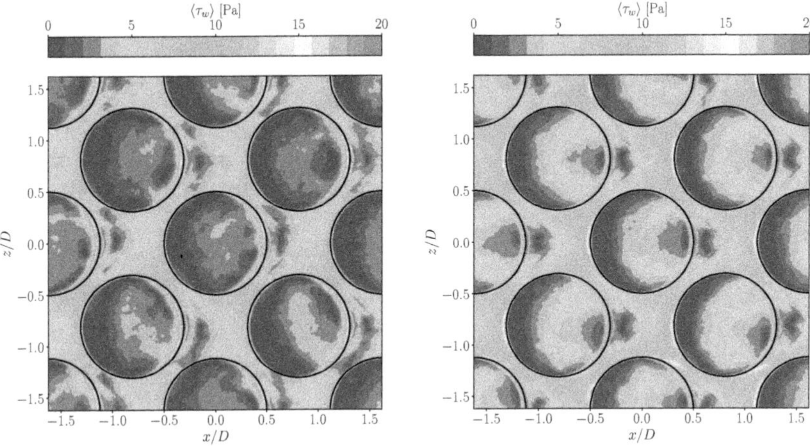

Fig. 7.28: Time-averaged wall shear stress distribution $\langle\tau_w\rangle$ at the clean, dimpled bottom wall at $Re_D = 42{,}000$: $t/D = 0.26$ (left) and $t/D = 0.35$ (right).

The reason for this observation is the relatively low channel height-to-dimple depth ratio of $H/D = 0.326$, since the vortex structures, generated by the dimples, are very large with respect to the restricted channel height. This results in stronger perturbations, higher turbulent mixing and a significant acceleration of the mean fluid flow. Consequently, higher time-averaged local wall shear stresses $\langle\tau_w\rangle$ can be observed at the bottom wall between the dimples and at the trailing edges, which have a substantial influence on the fouling distributions presented in the following section 7.2.3.

7.2.3 Prediction of particulate fouling

Fig. 7.29 shows the fouling layer height x_f and distribution after 150 s of physical real-time and a foulant particle mass loading $\eta = 0.2\%$. Opposed to the single spherical dimple (see Figs. 7.13 and 7.16), the simulated fouling layers for both dimple package configurations are very disturbed and not evenly distributed. The main reason for this observation is the high turbulence intensity and mixing, produced and enhanced by the dimples, which acts as turbulence generators. However, an identification of typical regions in the fouling layer is possible, for example the relatively high fouling layer thickness inside the recirculation zones, caused by lower fluid velocities and therefore reduced local wall shear stresses. The fouling layer thickness in this particular region is up to $x_f = 100\,\mu$m for the deep dimples and about $x_f = 80\,\mu$m for the shallow dimples. It is assumed, that a larger amount of the relatively heavy glass particles is not able to leave the deeper dimples (see also section 7.1.4), resulting in a higher amount of fouling deposits. In contrast, the high wall shear stresses due to the impingement of the high-speed mainstream flow in the back portion of the dimples mitigate particulate fouling. Similarly, the wall shear

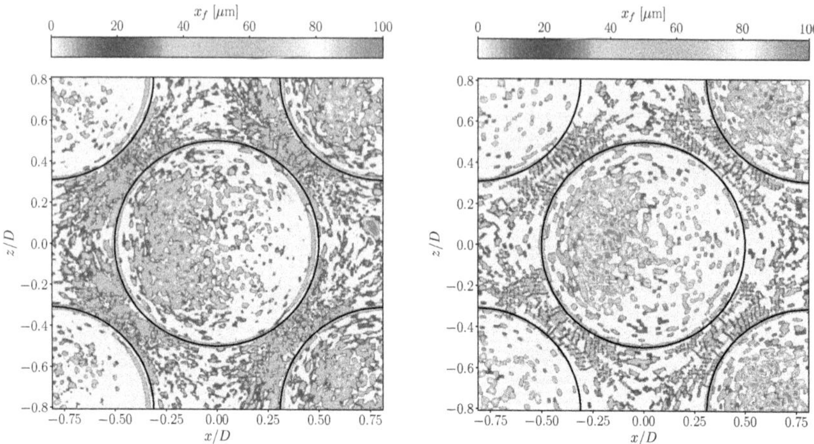

Fig. 7.29: Distribution and thickness x_f of fouling layer after 120 s of physical real time for a foulant particle mass loading $\eta = 0.2\%$ and $Re_D = 42{,}000$: $t/D = 0.26$ (left) and $t/D = 0.35$ (right).

stresses avoid the build up of evenly distributed fouling layers with large thicknesses on the remaining bottom wall, outside of the dimples. Solely for the shallow dimples, more particulate fouling occurs in the web region and in front of the dimples, amplified by lower fluid velocities and a stronger stagnation of the mean flow upstream of the dimples. This leads to a slightly higher overall amount of particulate fouling in case of the shallow dimples, which is contradictory to previous observations for the single spherical dimple, as presented in section 7.1.4 and 7.1.5. However, these findings can be explained by the completely different flow features, generated by a single dimple compared to several dimples in a staggered arrangement. Since a noticeable influence of the particulate fouling on the hydraulic losses and heat transfer could not be identified, the mass-based fouling resistance $R_{f,m}$ is used to assess the self-cleaning process induced by the investigated dimple packages. The drawback of this assessment is that the settled fouling mass m_f is recorded for the entire domain instead for local regions of interest. Thus, the presented mass-based fouling resistance describes the fouling behavior for the complete system as integral value. However, Fig. 7.30 shows the comparison between the asymptotic mass-based fouling resistance $R^*_{f,m} = 6.85 \times 10^{-5}\,\mathrm{m^2K/W}$ for the corresponding smooth channel configuration (i.e., similar fully turbulent channel flow without spherical dimples at the lower channel wall) and the progression of the mass-based fouling resistance in time, approaching its corresponding asymptotic value after approx. 120 s of physical real-time, for the shallow ($t/D = 0.26$) and deep dimple ($t/D = 0.35$) package. In both cases, a remarkable reduction of about 58% for $t/D = 0.26$ and of the (integral) mass-based fouling resistance is observed, which is about $\Delta R_{f,m} = 58\%$ for the shallow dimple configuration and about $\Delta R_{f,m} = 65\%$ for the deep dimple configuration, respectively. This observation is generally in line with predicted fouling layer distribution given in Fig. 7.29. Although the fouling behavior of dimple packages needs further investigations, the presented results provides a promising opportunity to mitigate particulate fouling in industrial applications (e.g., in heat exchanger or chiller units).

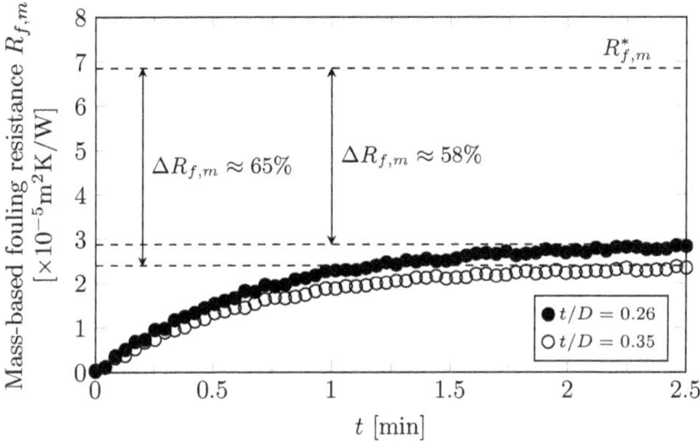

Fig. 7.30: Progression of the mass-based fouling resistance $R_{f,m}$ in time obtained for spherical dimples in a staggered arrangement with a dimple depth-to-dimple diameter ratio of $t/D = 0.26$ and $t/D = 0.35$.

7.2.4 Summary

The thermo-hydraulic performance analysis and prediction of particulate fouling for a staggered arrangement of deep dimples ($t/D = 0.35$) in a fully turbulent channel flow reveals the generation of local vortex structures which are larger and stronger than those induced by shallow dimples ($t/D = 0.26$) in a similar staggered arrangement. In general, these vortices promote stronger perturbations, causing a higher turbulent mixing as well as a significant acceleration of the mainstream fluid flow and consequently higher wall shear stresses. These locally enhanced wall shear stresses avoid the build up of fouling formations on the lower channel wall between the dimples. A slightly lower overall amount of particulate fouling and therefore a higher reduction of the mass-based fouling resistance of about $\Delta R_{f,m} = 65\%$ can be observed for the deep dimple package, which is contradictory to our previous observations for the single spherical dimple. However, this opposite finding is explained by completely different flow features generated by a single dimple in comparison with a dimple package.

8 Conclusion

This work aims to develop and to validate a numerical approach allowing spatial- and time-resolved simulations of particulate fouling on structured heat transfer surfaces, employable to investigate dimpled heat transfer surfaces regarding their ability to mitigate particulate fouling. In order to identify a possible self-cleaning process induced by dimples, a novel multiphase Eulerian-Lagrangian approach was formulated and implemented into the CFD framework "OpenFOAM". This method combines scale-resolving large-eddy simulations (LES) with an efficient Lagrangian point-particle tracking (LPT) algorithm, suitable not only to account for simple one-way momentum coupling, but also for two- and four-way phase coupling. The efficiency of the proposed fouling approach is additionally enhanced through a conversion algorithm which converts deposited particles into an additional continuous fouling phase (i.e., the fouling layer), followed by a deletion of these particles from the LPT. The fouling layer is treated as porous medium and accounts for increased hydraulic losses and for the reduced heat transfer (or fouling resistance). Additionally, a multiscale approch, based on the heterogeneous multiscale method (HMM), is suggested to overcome the high computational effort of multiphase Eulerian-Lagrangian LES for long-term fouling intervals. A comprehensive validation of the implemented numerical methods was carried out for different canonical test cases available in the literature (e.g., particle-laden flow in a simplified combustion chamber, particle-laden turbulent channel flow including particle deposition), ensuring the proper functionality and high accuracy of the proposed Eulerian-Lagrangian approach.

The investigation of different clean structured heat transfer surfaces, namely the single spherical dimple ($t/D = 0.26$ and 0.35) and a square cavity, revealed the superior thermo-hydraulic efficiency of dimpled surfaces mainly due to the impingement of the high-speed mainstream flow in the back portion of the dimple and the resulting upwash flow, which greatly enhances the turbulent mixing between the coolant flow and the hot flow in the near-wall region. In case of the soiled structured heat transfer surfaces, the oscillating mono-core vortex inside the dimple cavities could be identified as main reason for a self-cleaning process induced through dimpled surfaces, since the streamwise vorticity generated by this oscillating vortex structure leads to a local enhancement of the wall shear stresses and therefore to reduced particulate fouling. Moreover, the predicted fouling layer distributions disclosed a higher fouling-mitigation potential in case of the shallow dimple, which could by also confirmed by experimental fouling investigations. The main reason for superior fouling mitigation potential of the shallow dimple can be explained by the location of the high-vorticity regions, which are located closer to the lower channel wall, compared to the deep dimple configuration. The presented investigation for a staggered arrangement of deep dimples ($t/D = 0.35$) in a fully turbulent channel flow reveals the generation of local vortex structures which are larger and stronger than those induced by shallow dimples ($t/D = 0.26$) in a similar staggered arrangement. These vortices promote stronger perturbations, causing a higher turbulent mixing as well as a significant acceleration of the mainstream fluid flow and consequently higher wall shear stresses, avoiding the build up of fouling formations. A lower overall amount of fouling mass and thus a higher reduction of the mass-based fouling resistance could be observed for the deep dimple package, which is contradictory to the observations for the single spherical dimple. This can be explained by the completely different flow features generated by a single dimple in comparison to a dimple packages.

To sum it up, on the basis of the presented results, the supposed self-cleaning process induced by dimples could be identified and quantified for different single dimple configurations as well as for spherical dimples in a staggered arrangement, i.e., for dimple packages. The proposed multiphase Eulerian-Lagrangian approach generally allows a detailed and thorough numerical investigation of the interaction between local flow structures and particulate fouling for arbitrary structured heat transfer surfaces. Although its universality, the proposed method is not appropriate for the simulation of fouling processes within entire heat exhangers due to the high computational costs, but can greatly support the design process of innovate types of structured heat transfer surfaces with a strong immunity against particle depositions. Moreover, the employment of dimpled surfaces in industrial application provides a promising opportunity to achieve the best thermo-hydraulic efficiency while minimizing the susceptibility regarding particulate fouling.

Bibliography

[1] A. Abdulle, W. E, B. Engquist, and E. Vanden-Eijnden. The heterogeneous multiscale method. *Acta Numerica*, 21:1–87, 2012. doi:10.1017/s0962492912000025.

[2] V. N. Afanasyev, Y. P. Chudnovsky, A. I. Leontiev, and P. S. Roganov. Turbulent flow friction and heat transfer characteristics for spherical cavities on a flat plate. *Exp. Therm Fluid Sci.*, 7(1):1–8, 1993. doi:10.1016/0894-1777(93)90075-T.

[3] M. Alletto and M. Breuer. One-way, two-way and four-way coupled LES predictions of a particle-laden turbulent flow at high mass loading downstream of a confined bluff body. *Int. J. Multiphase Flow*, 45:70–90, 2012. doi:10.1016/j.ijmultiphaseflow.2012.05.005.

[4] F. Ambrosino. *Wall effects in particle-laden flows.* PhD thesis, University of Naples Federico II, 2011.

[5] B. Andersson, R. Andersson, L. Håkansson, M. Mortensen, R. Sudiyo, and B. van Wachem. *Computational Fluid Dynamics for Engineers.* Cambridge University Press, 2014.

[6] M. J. Andrews and P. J. O'Rourke. The multiphase particle-in-cell MP-PIC method for dense particulate flows. *Int. J. Multiphase Flow*, 22(2):379–402, 1996. doi:10.1016/0301-9322(95)00072-0.

[7] S. V. Apte, K. Mahesh, P. Moin, and J. C. Oefelein. Large-eddy simulation of swirling particle-laden flows in a coaxial-jet combustor. *Int. J. Multiphase Flow*, 29(8):1311–1331, 2003. doi:10.1016/s0301-9322(03)00104-6.

[8] R. Aris. *Vectors, Tensors and the Basic Equations of Fluid Mechanics.* Dover Publications Inc., 1990.

[9] K. Aroonrat and S. Wongwises. Experimental investigation of condensation heat transfer and pressure drop of R-134a flowing inside dimpled tubes with different dimpled depths. *Int. J. Heat Mass Transfer*, 128:783–793, 2019. doi:10.1016/j.ijheatmasstransfer.2018.09.039.

[10] C. Bailly and G. Comte-Bellot. *Turbulence.* Springer International Publishing, 2015. doi:10.1007/978-3-319-16160-0.

[11] A. B. Basset. III. On the motion of a sphere in a viscous liquid. *Philos. Trans. R. Soc. A*, 179:43–63, 1888. doi:https://doi.org/10.1098/rsta.1888.0003.

[12] F. Bianco, S. Chibbaro, C. Marchioli, M. V. Salvetti, and A. Soldati. Intrinsic filtering errors of Lagrangian particle tracking in LES flow fields. *Phys. Fluids*, 24:045103, 2012. doi:10.1063/1.3701378.

[13] R. Blöchl and H. M. Müller-Steinhagen. Influence of particle size and particle/fluid combination on particulate fouling in heat exchangers. *Can. J. Chem. Eng.*, 68(4):585–591, 1990. doi:10.1002/cjce.5450680408.

[14] U. Bobe. *Die Reinigbarkeit technischer Oberflächen im immergierten System.* PhD thesis, Technische Universität München, 2008. URL: http://mediatum.ub.tum.de/?id=649731.

[15] M. Bohnet. Fouling of Heat Transfer Surfaces. *Chem. Eng. Technol.*, 10(1):113–125, 1987. doi:10.1002/ceat.270100115.

[16] J. Borée, T. Ishima, and I. Flour. The effect of mass loading and inter-particle collisions on the development of the polydispersed two-phase flow downstream of a confined bluff body. *J. Fluid Mech.*, 443:129–165, 2001. doi:10.1017/S0022112001005134.

[17] T. R. Bott. *Fouling of Heat Exchangers.* Elsevier Science & Technology Books, 1995.

[18] J. V. Boussinesq. Sur la resistance d'une sphere solide. *C. R. Acad. Sci.*, 100:935, 1885.

[19] M. Breuer and M. Alletto. Efficient simulation of particle-laden turbulent flows with high mass loadings using LES. *Int. J. Heat Fluid Flow*, 35:2–12, 2012. doi:10.1016/j.ijheatfluidflow.2012.01.001.

[20] M. Breuer and W. Rodi. Large-Eddy Simulation of Turbulent Flow through a Straight Square Duct and a 180° Bend. *Fluid Mechanics and Its Applications*, 26:273–285, 1994. doi:https://doi.org/10.1007/978-94-011-1000-6_24.

[21] R. Bruchsal. *Der Einfluß von Partikelstoß und Partikelhaftung auf die Abscheidung in Faserfiltern.* PhD thesis, Technische Universität Karlsruhe, 1981.

[22] M. Chrigui, A. Hidouri, A. Sadiki, and J. Janicka. Unsteady Euler/Lagrange simulation of a confined bluff-body gas-solid turbulent flow. *Fluid Dyn. Res.*, 45(5):055501, 2013. doi:10.1088/0169-5983/45/5/055501.

[23] M. K. Chyu, Y. Yu, H. Ding, J. P. Downs, and F. O. Soechting. Concavity enhanced heat transfer in an internal cooling passage. In *Volume 3: Heat Transfer; Electric Power; Industrial and Cogeneration.* American Society of Mechanical Engineers, 1997. doi:10.1115/97-gt-437.

[24] D. P. Combest, P. A. Ramachandran, and M. P. Dudukovic. On the Gradient Diffusion Hypothesis and Passive Scalar Transport in Turbulent Flows. *Ind. Eng. Chem. Res.*, 50:8817–8823, 2011. doi:10.1021/ie200055s.

[25] C. T. Crowe, J. D. Schwarzkopf, M. Sommerfeld, and Y. Tsuji. *Multiphase Flows with Droplets and Particles.* CRC Press, 2011.

[26] C. T. Crowe, M. P. Sharma, and D. E. Stock. The Particle-Source-In Cell (PSI-CELL) Model for Gas-Droplet Flows. *J. Fluids Eng.*, 99(2), 1977. doi:10.1115/1.3448756.

[27] P. A. Cundall and O. D. L. Strack. A discrete numerical model for granular assemblies. *Géotechnique*, 29(1):47–65, 1979. doi:https://doi.org/10.1680/geot.1979.29.1.47.

[28] P. A. Davidson. *Turbulence - An Introduction for Scientists and Engineers.* Oxford University Press, 2004.

[29] M. De Marchis and B. Milici. Turbulence modulation by micro-particles in smooth and rough channels. *Phys. Fluids*, 28:115101, 2016. doi:10.1063/1.4966647.

[30] E. de Villiers. *The Potential of Large Eddy Simulation for theModeling of Wall Bounded Flows.* PhD thesis, Imperial College of Science, Technology and Medicine, 2006.

[31] R. B. Dean. Reynolds Number Dependence of Skin Friction and Other Bulk Flow Variables in Two-Dimensional Rectangular Duct Flow. *J. Fluids Eng*, 100(2):215–223, 1978. doi:10.1115/1.3448633.

[32] J. W. Deardorff. A numerical study of three-dimensional turbulent channel flow at large Reynolds numbers. *J. Fluid Mech.*, 41(2):453–480, apr 1970. doi:10.1017/s0022112070000691.

[33] H. Deponte, W. Augustin, and S. Scholl. Development of a quantification method for fouling deposits using phosphorescence. In *Proceedings of Heat Exchanger Fouling and Cleaning XIII*, pages 152–159, 2019.

[34] H. Deponte, R. Kasper, S. Schulte, W. Augustin, J. Turnow, N. Kornev, and S. Scholl. Experimental and numerical approach to resolve particle deposition on dimpled heat transfer surfaces locally and temporally. *Chem. Eng. Sci.*, 227:115840, 2020. doi:10.1016/j.ces.2020.115840.

[35] H. Deponte, L. Rohwer, W. Augustin, and S. Scholl. Investigation of deposition and self-cleaning mechanism during particulate fouling on dimpled surfaces. *Heat Mass Transfer*, 55(12):3633–3644, 2019. doi:10.1007/s00231-019-02676-0.

[36] B. Derjaguin and L. Landau. Theory of the stability of strongly charged lyophobic sols and of the adhesion of strongly charged particles in solutions of electrolytes. *Prog. Surf. Sci.*, 43(1–4):30–59, 1993. doi:10.1016/0079-6816(93)90013-L.

[37] W. E. *Principles of Multiscale Modeling.* Cambridge University Press, 2011.

[38] W. E, B. Engquist, and Z. Huang. Heterogeneous multiscale method: A general methodology for multiscale modeling. *Physical Review B*, 67(9), 2003. doi:10.1103/physrevb.67.092101.

[39] J. K. Eaton. Two-way coupled turbulence simulations of gas-particle flows using point-particle tracking. *Int. J. Multiphase Flow*, 35:792–800, 2009. doi:10.1016/j.ijmultiphaseflow.2009.02.009.

[40] H. Eckelmann. The structure of the viscous sublayer and the adjacent wall region in a turbulent channel flow. *J. Fluid Mech.*, 65(3):439–459, 1974. doi:https://doi.org/10.1017/S0022112074001479.

[41] S. Elghobashi. On predicting particle-laden turbulent flows. *Appl. Sci. Res.*, 52(4):309–329, 1993. doi:10.1007/BF00936835.

[42] M. A. Elyyan, A. Rozati, and D. K. Tafti. Investigation of dimpled fins for heat transfer enhancement in compact heat exchangers. *Int. J. Heat Mass Transfer*, 51(11–12):2950–2966, 2008. doi:10.1016/j.ijheatmasstransfer.2007.09.013.

[43] N. Epstein. Fouling in heat exchangers. *6th Intern. Heat Transf. Conf., Toronto*, pages 235–253, 1978.

[44] N. Epstein. Thinking about Heat Transfer Fouling: A 5×5 Matrix. *Heat Transfer Eng.*, 4(1):43–56, 1983. doi:10.1080/01457638108939594.

[45] N. Epstein. Fouling Science and Technology: Particulate Fouling. *NATO ASI Series*, 145:143–164, 1988.

[46] J. H. Ferziger, M. Perić, and R. L. Street. *Computational Methods for Fluid Dynamics*. Springer International Publishing, 2020. doi:10.1007/978-3-319-99693-6.

[47] J. R. Fessler and J. K. Eaton. Turbulence modification by particles in a backward-facing step flow. *J. Fluid Mech.*, 394:97–117, 1999. doi:10.1017/S0022112099005741.

[48] J. Fröhlich. *Large Eddy Simulation turbulenter Strömungen*. Teubner, 2006. doi:10.1007/978-3-8351-9051-1.

[49] D. L. Gee and R. L. Webb. Forced Convection Heat Transfer in Helically Rib-Roughened Tubes. *Int. J. Heat Mass Transfer*, 23(8):1127–1136, 1980. doi:10.1016/0017-9310(80)90177-5.

[50] M. Germano. A proposal for a redefinition of the turbulent stresses in the filtered Navier-Stokes equations. *Phys. Fluids*, 29(7):2323, 1986. doi:10.1063/1.865568.

[51] M. Germano. Turbulence: the filtering approach. *J. Fluid Mech.*, 238:325–336, 1992. doi:https://doi.org/10.1017/S0022112092001733.

[52] M. Germano, U. Piomelli, and W. H. Cabot. A dynamic subgrid-scale eddy viscosity model. *Phys. Fluids*, 3(3):1760–1765, 1991. doi:10.1063/1.857955.

[53] K. Gersten and H. Herwig. *Strömungsmechanik*. Vieweg+Teubner Verlag, 1992. doi:https://doi.org/10.1007/978-3-322-93970-8.

[54] S. Ghosal and P. Moin. The Basic Equations for the Large Eddy Simulation of Turbulent Flows in Complex Geometry. *J. Comput. Phys.*, 118(1):24–37, 1995. doi:10.1006/jcph.1995.1077.

[55] A. Haider and O. Levenspiel. Drag coefficient and terminal velocity of spherical and nonspherical particles. *Powder Technol.*, 58(1):63–70, 1989. doi:10.1016/0032-5910(89)80008-7.

[56] E. Heinl and M. Bohnet. Calculation of particle-wall adhesion in horizontal gas-solids flow using CFD. *Powder Technol.*, 159(2):95–104, 2005. doi:10.1016/j.powtec.2004.09.037.

[57] C. Henry, J.-P. Minier, and G. Lefèvre. Towards a description of particulate fouling: From single particle deposition to clogging. *Adv. Colloid Interface Sci.*, 185-186:34–76, 2012. doi:10.1016/j.cis.2012.10.001.

[58] J. O. Hinze. *Turbulence*. McGraw-Hill Inc., 1975.

[59] C. Hirsch. *Numerical Computation of Internal and External Flows - The Fundamentals of Computational Fluid Dynamics*. Butterworth-Heinemann, 2007. doi:10.1016/B978-0-7506-6594-0.X5037-1.

[60] J. J. H. Houben. *Experimental Investigations and CFD Simulations onParticle Depositions in Gas Cyclone Separators*. PhD thesis, Montanuniversität Leoben, 2011.

[61] F. P. Incropera, D. P. DeWitt, T. L. Bergman, and A. S. Lavine. *Fundamentals of Heat and Mass Transfer*. John Wiley & Sons, 2006.

[62] S. A. Isaev, N. V. Kornev, A. I. Leontiev, and E. Hassel. Influence of the Reynolds number and the spherical dimple depth on turbulent heat transfer and hydraulic loss in a narrow channel. *Int. J. Heat Mass Transfer*, 53(1–3):178–197, 2010. doi:10.1016/j.ijheatmasstransfer.2009.09.042.

[63] R. I. Issa. Solution of the implicitly discretised fluid flow equations by operator-splitting. *J. Comput. Phys.*, 62(1):40–65, 1986. doi:10.1016/0021-9991(86)90099-9.

[64] H. Jasak. *Error Analysis and Estimation for the Finite Volume Method with Applications to Fluid Flows*. PhD thesis, Department of Mechanical Engineering, Imperial College, 1996.

[65] B. A. Kader. Temperature and concentration profiles in fully turbulent boundary layers. *Int. J. Heat Mass Transfer*, 24(9):1541–1544, 1981. doi:10.1016/0017-9310(81)90220-9.

[66] R. Kasper, H. Deponte, A. Michel, J. Turnow, W. Augustin, S. Scholl, and N. Kornev. Numerical investigation of the interaction between local flow structures and particulate fouling on structured heat transfer surfaces. *Int. J. Heat Fluid Flow*, 71:68–79, 2018. doi:10.1016/j.ijheatfluidflow.2018.03.002.

[67] R. Kasper, J. Turnow, and N. Kornev. Numerical modeling and simulation of particulate fouling of structured heat transfer surfaces using a multiphase Euler-Lagrange approach. *Int. J. Heat Mass Transfer*, 115:932–945, 2017. doi:10.1016/j.ijheatmasstransfer.2017.07.108.

[68] R. Kasper, J. Turnow, and N. Kornev. Multiphase Eulerian-Lagrangian LES of particulate fouling on structured heat transfer surfaces. *Int. J. Heat Fluid Flow*, 79:108462, 2019. doi:10.1016/j.ijheatfluidflow.2019.108462.

[69] D. Q. Kern and R. A. Seaton. A Theoretical Analysis of Thermal Surface Fouling. *Brit. Chem. Eng.*, 4(5):258–262, 1959.

[70] J. Kim. On the structure of pressure fluctuations in simulated turbulent channel flow. *J. Fluid Mech.*, 205:421–451, 1989. doi:https://doi.org/10.1017/S0022112089002090.

[71] J. Kim, P. Moin, and R. Moser. Turbulence statistics in fully developed channel flow at low Reynolds number. *J. Fluid Mech.*, 177:133–166, 1987. doi:https://doi.org/10.1017/S0022112087000892.

[72] N. A. Kiselev, A. I. Leontiev, Y. A. Vinogradov, A. G. Zditovets, and M. M. Strongin. Effect of large-scale vortex induced by a cylinder on the drag and heat transfer coefficients of smooth and dimpled surfaces. *Int. J. Therm. Sci.*, 136:396–409, 2019. doi:10.1016/j.ijthermalsci.2018.11.005.

[73] A. N. Kolmogorov. The local structure of turbulence in incompressible viscous fluid for very large Reynolds numbers. *Dokl. Akad. Nauk SSSR*, 30:299–303, 1941.

[74] N. Kornev, I. Shevchuk, N. Abbas, P. Anschau, and S. Samarbakhsh. Potential and limitations of scale resolved simulations for ship hydrodynamics applications. *Ship Technology Research*, 66(2):83–96, 2019. doi:https://doi.org/10.1080/09377255.2019.1574965.

[75] A. Kozlov and Y. Chudnovsky. Experimental evaluation of heat transfer, pressure drop, and fouling mitigation potential in finned, dimpled, and bare tube bundles. *Int. J. Energy Clean Environ.*, 9(1-3):103–118, 2008. doi:10.1615/interjenercleanenv.v9.i1-3.80.

[76] A. G. Kravchenko, H. Choi, and P. Moin. On the relation of near-wall streamwise vortices to wall skin friction in turbulent boundary layers. *Phys. Fluids A*, 5(12):3307–3309, 1993. doi:10.1063/1.858692.

[77] O. Krischer. *Die wissenschaftlichen Grundlagen der Trocknungstechnik*. Springer Berlin Heidelberg, 1956. doi:10.1007/978-3-662-26010-4.

[78] P. K. Kundu, I. M. Cohan, and D. R. Downling. *Fluid mechanics*. Academic Press, 6th edition, 2015.

[79] C. Kunik. *CFD-Simulation turbulenter konvektiver Strömungen bei überkritischen Drücken*. PhD thesis, Karlsruher Institut für Technologie (KIT), 2012.

[80] B. Launder and N. Sandham, editors. *Closure Strategies for Turbulent and Transitional Flows.* Cambridge University Press, 2002.

[81] A. Leonard. Energy Cascade in Large-Eddy Simulations of Turbulent Fluid Flows. In *Turbulent Diffusion in Environmental Pollution, Proceedings of a Symposium held at Charlottesville*, pages 237–248. Elsevier, 1975. doi:10.1016/s0065-2687(08)60464-1.

[82] M. Lesieur. *Turbulence in Fluids.* Springer Netherlands, 2008. doi:10.1007/978-1-4020-6435-7.

[83] P. M. Ligrani. Heat Transfer Augmentation Technologies for Internal Cooling of Turbine Components of Gas Turbine Engines. *Int. J. Rotating Mach.*, 2013:1–32, 2013. doi:10.1155/2013/275653.

[84] P. M. Ligrani, T. Blaskovich, and M. M. Oliveira. Comparison of Heat Transfer Augmentation Techniques. *AIAA Journal*, 41(3):337–362, 2003. doi:10.2514/2.1964.

[85] P. M. Ligrani, J. L. Harrison, G. I. Mahmmod, and M. L. Hill. Flow structure due to dimple depressions on a channel surface. *Phys. Fluids*, 13(11):3442–3451, 2001. doi:10.1063/1.1404139.

[86] D. K. Lilly. A proposed modification of the Germano subgrid-scale closure method. *Phys. Fluids*, 4(3):633–635, 1992. doi:https://doi.org/10.1063/1.858280.

[87] Y. Liu, Y. Rao, and B. Weigand. Heat transfer and pressure loss characteristics in a swirl cooling tube with dimples on the tube inner surface. *Int. J. Heat Mass Transfer*, 128:54–65, 2018. doi:10.1016/j.ijheatmasstransfer.2018.08.097.

[88] J. Lui, Y. Song, G. Xie, and B. Sunden. Numerical modeling flow and heat transfer in dimpled cooling channels with secondary hemispherical protrusions. *Energy*, 79:1–19, 2015. doi:10.1016/j.energy.2014.05.075.

[89] J. L. Lumley and H. Tennekes. *A first course in turbulence.* The MIT Press, 1972.

[90] T. S. Lund. The use of explicit filters in large eddy simulation. *Comput. Math. Appl.*, 46(4):603–616, 2003. doi:10.1016/s0898-1221(03)90019-8.

[91] T. S. Lund, X. Wu, and K. D. Squires. Generation of Turbulent Inflow Data for Spatially-Developing Boundary Layer Simulations. *J. Comput. Phys.*, 140(2):233–258, 1998. doi:10.1006/jcph.1998.5882.

[92] K. Luo, Q. Dai, X. Liu, and J. Fan. Effects of wall roughness on particle dynamics in a spatially developing turbulentboundary layer. *Int. J. Multiphase Flow*, 111:140–157, 2019. doi:10.1016/j.ijmultiphaseflow.2018.11.015.

[93] G. B. Macpherson, N. Nordin, and H. G. Weller. Particle tracking in unstructured, arbitrary polyhedral meshes for use in CFD and molecular dynamics. *Commun. Numer. Methods Eng.*, 25(3):263–273, 2009. doi:10.1002/cnm.1128.

[94] G. B. Macpherson and J. M. Reese. Molecular dynamics in arbitrary geometries: Parallel evaluation of pair forces. *Mol. Simul.*, 34(1):97–115, 2008. doi:10.1080/08927020801930554.

[95] G. I. Mahmood, P. M. Ligrani, and M. Z. Sabbagh. Heat Transfer in a Channel with Dimples and Protrusions on Opposite Walls. *J. Thermophys Heat Transfer*, 15(3):275–283, 2001. doi:10.2514/2.6623.

[96] C. Marchioli. Large-eddy simulation of turbulent dispersed flows: a review of modelling approaches. *Acta Mech.*, 228:741–771, 2017. doi:10.1007/s00707-017-1803-x.

[97] C. Marchioli, V. Armenio, M. V. Salvetti, and A. Soldati. Mechanisms for deposition and resuspension of heavy particles in turbulent flow over wavy interfaces. *Phys. Fluids*, 18:025102, 2006.

[98] C. Marchioli, A. Giusti, M. V. Salvetti, and A. Soldati. Direct numerical simulation of particle wall transfer and deposition in upward turbulent pipe flow. *Int. J. Multiphase Flow*, 29(6):1017–1038, 2003. doi:10.1016/s0301-9322(03)00036-3.

[99] C. Marchioli, M. V. Salvetti, and A. Soldati. Some issues concerning large-eddy simulation of inertial particle dispersion in turbulent bounded flows. *Phys. Fluids*, 20:1–11, 2008. doi:10.1063/1.2911018.

[100] C. Marchioli, A. Soldati, J. G. M. Kuerten, B. Arcen, A. Tanière, G. Goldensoph, K. D. Squires, M. F. Cargnelutti, and L. M. Portela. Statistics of particle dispersion in direct numerical simulations of wall-bounded turbulence: Results of an international collaborative benchmark test. *Int. J. Multiphase Flow*, 34:879–893, 2008. doi:10.1016/j.ijmultiphaseflow.2008.01.009.

[101] M. R. Maxey and J. J. Riley. Equation of motion for a small rigid sphere in a nonuniform flow. *Phys. Fluids*, 26(4):883, 1983. doi:https://doi.org/10.1063/1.864230.

[102] C. Meneveau and J. Katz. Scale-Invariance and Turbulence Models for Large-Eddy Simulation. *Annu. Rev. Fluid Mech.*, 32(1):1–32, 2000. doi:10.1146/annurev.fluid.32.1.1.

[103] F. R. Menter. Two-equation eddy-viscosity turbulence models for engineering applications. *AIAA Journal*, 32(8):1598–1605, 1994. doi:10.2514/3.12149.

[104] E. E. Michaelides, C. T. Crowe, and J. D. Schwarzkopf, editors. *Multiphase Flow Handbook*. Taylor & Francis Inc, 2016.

[105] B. Milici, M. De Marchis, G. Sardina, and E. Napoli. Effects of roughness on particle dynamics inturbulent channel flows: a DNS analysis. *J. Fluid Mech.*, 739:465–478, 2014. doi:10.1017/jfm.2013.633.

[106] P. Moin and J. Kim. Numerical investigation of turbulent channel flow. *J. Fluid Mech.*, 118(1):341–377, 1982. doi:10.1017/s0022112082001116.

[107] R. D. Moser, J. Kim, and N. N. Mansour. Direct numerical simulation of turbulent channel flow up to $\mathrm{Re}_\tau = 590$. *Phys. Fluids*, 11(4):943–945, 1999. doi:10.1063/1.869966.

[108] H. M. Müller-Steinhagen. *VDI Heat Atlas - Fouling of Heat Exchanger Surfaces.* Springer-Verlag Berlin Heidelberg, 2010. doi:10.1007/978-3-540-77877-6.

[109] H. M. Müller-Steinhagen. Heat Transfer Fouling: 50 Years After the Kern and Seaton Model. *Heat Transfer Eng.*, 32(1):1–13, 2011. doi:10.1080/01457632.2010.505127.

[110] H. M. Müller-Steinhagen and J. Middis. Particulate Fouling in Plate Heat Exchangers. *Heat Transfer Eng.*, 10(4):30–36, jan 1989. doi:10.1080/01457638908939714.

[111] C. Narayanan, D. Lakehal, L. Botto, and A. Soldati. Mechanisms of particle deposition in a fully developed turbulent open channel flow. *Phys. Fluids*, 15(3):763–775, 2003. doi:10.1063/1.1545473.

[112] N. Nordin. *Complex chemistry modeling of diesel spray combustion.* PhD thesis, Chalmers University of Technology, 2000.

[113] C. W. Oseen. *Hydromechanik.* Akademische Verlagsgem. Leipzig, 1927.

[114] U. Piomelli, P. Moin, and J. H. Ferziger. Models for Large Eddy Simulations of Turbulent Channel Flows including Transpiration. *AIAA J. Thermophys. Heat Transf.*, 5:124–128, 1991.

[115] W. Polifke and J. Kopitz. *Wärmeübertragung.* Pearson Studium, 2009.

[116] S. B. Pope. *Turbulent Flows.* Cambridge University Press, 2000. doi:10.1017/cbo9780511840531.

[117] J. Pozorski and S. V. Apte. Filtered particle tracking in istropic turbulence and stochastic modeling subgrid-scale dispersion. *Int. J. Multiphase Flow*, 35(2):118–128, 2009. doi:10.1016/j.ijmultiphaseflow.2008.10.005.

[118] A. Putnam. Integrable form of droplet drag coefficient. *ARS Journal*, 31:1467–1470, 1961.

[119] M. W. Reeks. The transport of discrete particles in inhomogeneous turbulence. *J. Aerosol Sci.*, 14(6):729–739, 1983. doi:10.1016/0021-8502(83)90055-1.

[120] M. W. Reeks, J. Reed, and D. Hall. On the resuspension of small particles by a turbulent flow. *J. Phys. D: Appl. Phys.*, 21(4):574–589, 1988. doi:10.1088/0022-3727/21/4/006.

[121] O. Reynolds. IV. on the dynamical theory of incompressible viscous fluids and the determination of the criterion. *Philosophical Transactions of the Royal Society of London. (A.)*, 186:123–164, 1895. doi:10.1098/rsta.1895.0004.

[122] C. M. Rhie and W. L. Chow. Numerical study of the turbulent flow past an airfoil with trailing edge separation. *AIAA Journal*, 21(11):1525–1532, 1983. doi:10.2514/3.8284.

[123] P. Sagaut. *Large Eddy Simulation for Incompressible Flows*. Springer-Verlag Berlin Heidelberg, 3rd edition, 2006. doi:10.1007/b137536.

[124] P. Sagaut, S. Deck, and M. Terracol. *Multiscale and Multiresolution Approaches in Turbulence - LES, DES and Hybrid RANS/LES Methods*. Imperial College Press, 2013.

[125] G. Sardina, F. Picano, P. Schlatter, L. Brandt, and C. M. Casciola. Large Scale Accumulation Patterns of Inertial Particles in Wall-Bounded Turbulent Flow. *Flow Turbulence Combust*, 86:519–532, 2011. doi:10.1007/s10494-010-9322-z.

[126] P. Schlatter. *Large-Eddy simulation of transition and turbulence in wall-bounded shear flows*. PhD thesis, Swiss Federal Institute of Technology Zürich (ETH), 2005.

[127] H. Schlichting and K. Gersten. *Boundary-Layer Theory*. Springer Berlin Heidelberg, 9th edition, 2017. doi:10.1007/978-3-662-52919-5.

[128] G. A. Sehmel. Particle eddy diffusivities and deposition velocities for isothermal flow and smooth surfaces. *Aerosol Sci.*, 4(2):125–138, 1973. doi:10.1016/0021-8502(73)90064-5.

[129] M. L. Shur, P. R. Spalart, M. K. Strelets, and A. K. Travin. A hybrid RANS-LES approach with delayed-DES and wall-modelled LES capabilities. *Int. J. Heat Fluid Flow*, 29(6):1638–1649, 2008. doi:10.1016/j.ijheatfluidflow.2008.07.001.

[130] J. Smagorinsky. General circulation experiments with the primitive equations. *Mon. Weather Rev.*, 91(3):99–164, 1963. doi:10.1175/1520-0493(1963)091<0099:gcewtp>2.3.co;2.

[131] L. E. Smart and E. A. Moore. *Solid State Chemistry: An Introduction*. Taylor & Francis Inc, 2012.

[132] A. J. Smits, B. J. McKeon, and I. Marusic. High-Reynolds Number Wall Turbulence. *Annu. Rev. Fluid Mech.*, 43(1):353–375, 2011. doi:https://doi.org/10.1146/annurev-fluid-122109-160753.

[133] A. Soldati. Particles turbulence interactions in boundary layers. *Z. Angew. Math. Mech.*, 85(10), 2005. doi:10.1002/zamm.200410213.

[134] A. Soldati and C. Marchioli. Physics and modelling of turbulent particle deposition and entrainment: Review of a systematic study. *Int. J. Multiphase Flow*, 35(9):827–839, 2009. doi:10.1016/j.ijmultiphaseflow.2009.02.016.

[135] M. Sommerfeld. *VDI Heat Atlas - Particle Motion in Fluids*. Springer-Verlag Berlin Heidelberg, 2010. doi:10.1007/978-3-540-77877-6.

[136] M. Sommerfeld. *Particles in Flows*, chapter Numerical Methods for Dispersed Multiphase Flows, pages 327–396. Springer International Publishing, 2017. doi:10.1007/978-3-319-60282-0.

[137] D. B. Spalding. A Single Formula for the "Law of the Wall". *J. Appl. Mech.*, 28(3):455–458, 1961. doi:10.1115/1.3641728.

[138] C. G. Speziale. Galilean invariance of subgrid-scale stress models in the large-eddy simulation of turbulence. *J. Fluid Mech.*, 156:55, 1985. doi:10.1017/s0022112085001987.

[139] R. Steinhagen, H. Müller-Steinhagen, and K. Maani. Problems and Costs due to Heat Exchanger Fouling in New Zealand Industries. *Heat Transfer Eng.*, 14(1):19–30, 1993. doi:10.1080/01457639308939791.

[140] G. Stokes. On the Effect of the Internal Friction of Fluids on the Motion of Pendulums. In *Mathematical and Physical Papers*. Cambridge University Press, 2009. doi:10.1017/CBO9780511702266.002.

[141] J. W. Suitor, J. Marner, and R. B. Ritter. The history and status of research in fouling of heat exchanger in cooling water service. *Can. J. Chem. Eng.*, 55(8):374–380, 1977. doi:10.1002/cjce.5450550402.

[142] J. Taborek, T. Aoki, R. B. Ritter, J. W. Palen, and J. G. Knudsen. Predictive method for fouling behavior. *Chem. Eng. Prog.*, 68(7):69–78, 1972.

[143] V. I. Terekhov, S. V. Kalinina, and Y. M. Mshvidobadze. Heat transfer coefficient and aerodynamic resistance on a surface with a single dimple. *J. Enh. Heat Transf.*, 4(2):131–145, 1997. doi:10.1615/JEnhHeatTransf.v4.i2.60.

[144] Z.-X. Tong, M.-J. Li, Y.-L. He, and H.-Z. Tan. Simulation of real time particle deposition and removal processes on tubes by coupled numerical method. *Appl. Energy*, 185(2):2181–2193, 2017. doi:10.1016/j.apenergy.2016.01.043.

[145] Y. Tsuji, T. Tanaka, and T. Ishida. Lagrangian numerical simulation of plug flow of cohesionless particles in a horizontal pipe. *Powder Technol.*, 71(3):239–250, 1992. doi:10.1016/0032-5910(92)88030-L.

[146] J. Turnow. *Flow structures and heat transfer on dimpled surfaces*. PhD thesis, University of Rostock, 2012. URL: http://purl.uni-rostock.de/rosdok/id00001035.

[147] J. Turnow, R. Kasper, and N. Kornev. Flow structures and heat transfer over a single dimple using hybrid URANS-LES methods. *Comput. Fluids*, 172:720–727, 2018. doi:10.1016/j.compfluid.2018.01.014.

[148] J. Turnow, N. Kornev, S. Isaev, and E. Hassel. Vortex mechanism of heat transfer enhancement in a channel with spherical and oval dimples. *Heat Mass Transfer*, 47(3):301–313, 2011. doi:10.1007/s00231-010-0720-5.

[149] J. Turnow, N. Kornev, V. Zhdanov, and E. Hassel. Flow structures and heat transfer

on dimples in a staggered arrangement. *Int. J. Heat Fluid Flow*, 35:168–175, 2012. doi:0.1016/j.ijheatfluidflow.2012.01.002.

[150] E. R. van Driest. On Turbulent Flow Near a Wall. *J. Aeronaut. Sci.*, 23(11):1007–1011, 1956. doi:10.2514/8.3713.

[151] H. Versteeg and W. Malalasekera. *An Introduction to Computational Fluid Dynamics: The Finite Volume Method.* Prentice Hall, 2007.

[152] E. J. W. Verwey and J. T. G. Overbeek. *Theory of the Stability of Lyophobic Colloids.* Dover Pubn Inc, 1999.

[153] J. Visser. Van der Waals and other cohesive forces affecting powder fluidization. *Powder Technol.*, 58(1):1–10, 1989. doi:https://doi.org/10.1016/0032-5910(89)80001-4.

[154] J. Visser. Particle adhesion and removal: a review. *Part. Sci. Technol.*, 13:169–196, 1995. doi:10.1080/02726359508906677.

[155] N. Vorayos, N. Katkhaw, T. Kiatsiriroat, and A. Nuntaphan. Heat transfer behavior of flat plate having spherical dimpled surfaces. *Case Stud. Therm. Eng.*, 8:370–377, 2016. doi:10.1016/j.csite.2016.09.004.

[156] Ž. Tuković and H. Jasak. A moving mesh finite volume interface tracking method for surface tension dominated interfacial fluid flow. *Comput. Fluids*, 55:70–84, 2012. doi:10.1016/j.compfluid.2011.11.003.

[157] M. Walter. *Untersuchung der passiven und reaktiven Mischung in einem Strahlmischer bei hohen Reynolds und Schmidt Zahlen mittels Grobstruktursimulation.* PhD thesis, University of Rostock, 2017.

[158] F.-J. Wang, J.-X. Zhu, and J. M. Beeckmans. Pressure gradient and particle adhesion in the pneumatic transport of cohesive fine powders. *Int. J. Multiphase Flow*, 26:245–265, 2000. doi:10.1016/S0301-9322(99)00009-9.

[159] F.-L. Wang, Y.-L. He, S.-Z. Tang, and Z.-X. Tong. Parameter study on the fouling characteristics of the H-type finned tube heat exchangers. *Int. J. Heat Mass Transfer*, 112:367–378, 2017. doi:10.1016/j.ijheatmasstransfer.2017.04.107.

[160] F.-L. Wang, Y.-L. He, Z.-X. Tong, and S.-Z. Tang. Real-time fouling characteristics of a typical heat exchanger used in the waste heat recovery systems. *Int. J. Heat Mass Transfer*, 104:774–786, 2017. doi:10.1016/j.ijheatmasstransfer.2016.08.112.

[161] Q. Wang and K. D. Squires. Large eddy simulation of particle deposition in a vertical turbulent channel flow. *Int. J. Multiphase flow*, 22(4):667–683, 1996. doi:10.1016/0301-9322(96)00007-9.

[162] N. P. Waterson and H. Deconinck. Design principles for bounded higher-order convection schemes - a unified approach. *J. Comput. Phys.*, 224(1):182–207, 2007. doi:10.1016/j.jcp.2007.01.021.

[163] P. Welch. The use of fast Fourier transform for the estimation of power spectra: A method based on time averaging over short, modified periodograms. *IEEE Transactions on Audio and Electroacoustics*, 15(2):70–73, 1967. doi:10.1109/tau.1967.1161901.

[164] H. Y. Wen and G. Kasper. On the kinetics of particle reentrainment from surfaces. *J. Aerosol Sci.*, 20(4):483–498, 1989. doi:10.1016/0021-8502(89)90082-7.

[165] S. Wetchagarun. *A Numerical Study of Turbulent Two-phase Flows.* PhD thesis, University of Washington, 2008.

[166] D. C. Wilcox. *Turbulence Modeling for CFD.* DCW Industries Inc., 3rd edition, 2006.

[167] A. Yoshizawa. A statistically-derived subgrid model for the large-eddy simulation of turbulence. *Phys, Fluids*, 25(9):1532–1538, 1982. doi:10.1063/1.863940.

[168] A. Yoshizawa and K. Horiuti. A Statistically-Derived Subgrid-Scale Kinetic Energy Model for the Large-Eddy Simulation of Turbulent Flows. *J. Phys. Soc. Jpn.*, 54(8):2834–2839, 1985. doi:10.1143/JPSJ.54.2834.

[169] H. Zhang and G. Ahmadi. Aerosol particle transport and deposition in vertical and horizontal turbulent duct flows. *J. Fluid Mech.*, 406:55–80, 2000. doi:10.1017/S0022112099007284.

[170] J. Zhang and A. Li. CFD simulation of particle deposition in a horizontal turbulent duct flow. *Chem. Eng. Res. Des.*, 86(1):95–106, 2008. doi:10.1016/j.cherd.2007.10.014.

[171] F. Zhao and B. G. M. van Wachem. Direct numerical simulation of ellipsoidal particles in turbulent channel flow. *Acta Mech*, 224:2331–2358, 2013. doi:10.1007/s00707-013-0921-3.

[172] W. Zhou, Y. Rao, and H. Hu. An Experimental Investigation on the Characteristics of Turbulent Boundary Layer Flows Over a Dimpled Surface. *J. Fluids Eng.*, 138(2), 2015. doi:10.1115/1.4031260.

[173] G. Ziskind, M. Fichman, and C. Gutfinger. Resuspension of particulates from surfaces to turbulent flows—review and analysis. *J. Aerosol Sci.*, 26(4):613–644, 1995. doi:10.1016/0021-8502(94)00139-p.

A Appendix

A.1 High order statistical moments

The standard deviation σ, skewness S and flatness F (also referred to as kurtosis) of any arbitrary flow variable ϕ is defined as [116]:

$$\sigma_\phi = \sqrt{\mu_2(\phi)}, \tag{A.1}$$

$$S_\phi = \frac{\mu_3(\phi)}{\langle \phi'^2 \rangle^{3/2}}, \tag{A.2}$$

$$F_\phi = \frac{\mu_4(\phi)}{\langle \phi'^2 \rangle^{2}}, \tag{A.3}$$

based on the second, third and forth (central) statistical moments, respectively, which are evaluated as follows:

$$\mu_2(\phi) = \langle \phi'^2 \rangle = \langle (\phi - \langle \phi \rangle)^2 \rangle = \langle \phi^2 \rangle - \langle \phi \rangle^2, \tag{A.4}$$

$$\mu_3(\phi) = \langle \phi'^3 \rangle = \langle \phi^3 \rangle - 3\langle \phi \rangle \langle \phi^2 \rangle + 2\langle \phi \rangle^3, \tag{A.5}$$

$$\mu_4(\phi) = \langle \phi'^4 \rangle = \langle \phi^4 \rangle - 4\langle \phi \rangle \langle \phi^3 \rangle + 6\langle \phi \rangle^2 \langle \phi^2 \rangle - 3\langle \phi \rangle^4. \tag{A.6}$$

The variance σ_ϕ describes the deviation from its mean value $\langle \phi \rangle$. The skewness S_ϕ is a measure for the asymmetry of the probability distribution of ϕ, whereas the flatness F_ϕ is a measure for the curvature or slope of the probability distribution of ϕ.

A.2 Unladen turbulent channel flow over a single spherical dimple

This section presents supplementary results for the unladen turbulent channel flow over a single spherical dimple with a dimple depth-to-dimple diameter ratio of $t/D = 0.26$ and $t/D = 0.35$. This includes for example a grid independence study (GIS) and shows the influence of different subgrid-scale (SGS) models (Smagorinsky model (SM), dynamic Smagorinsky model (DSM) and dynamic one equation eddy-viscosity model (DOE)) as well as the Reynolds number Re_D (i.e., the average bulk flow velocity) on the flow fields.

A.2.1 Grid independence study

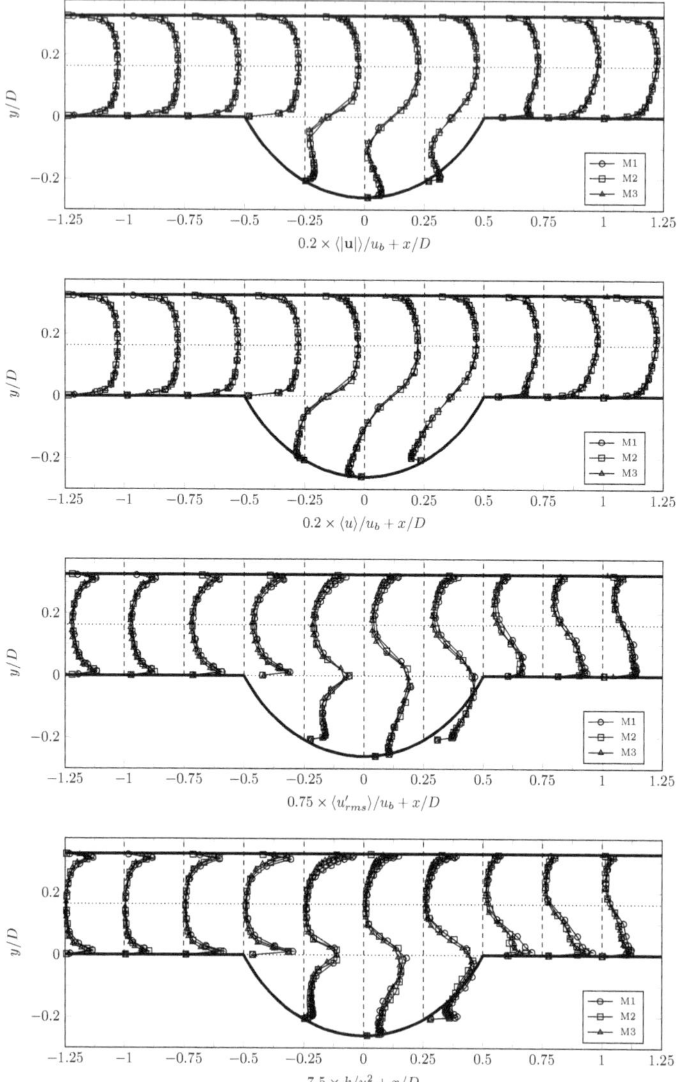

Fig. A.1: Grid independence study for a unladen turbulent channel flow over a single spherical dimple with a dimple depth-to-dimple diameter ratio of $t/D = 0.26$ at $Re_D = 42{,}000$.

A.2.2 Influence of the subgrid-scale model

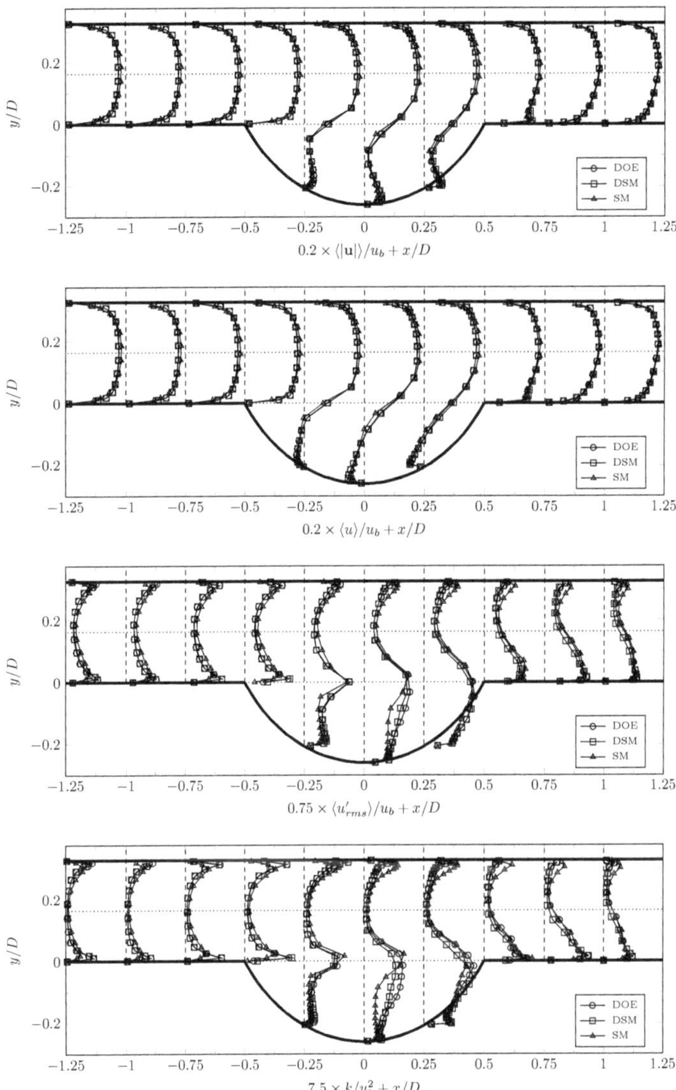

Fig. A.2: Influence of the subgrid-scale model on the unladen turbulent channel flow over a single spherical dimple with a dimple depth-to-dimple diameter ratio of $t/D = 0.26$ at $Re_D = 42{,}000$.

A.2.3 Influence of the Reynolds number ($t/D = 0.26$)

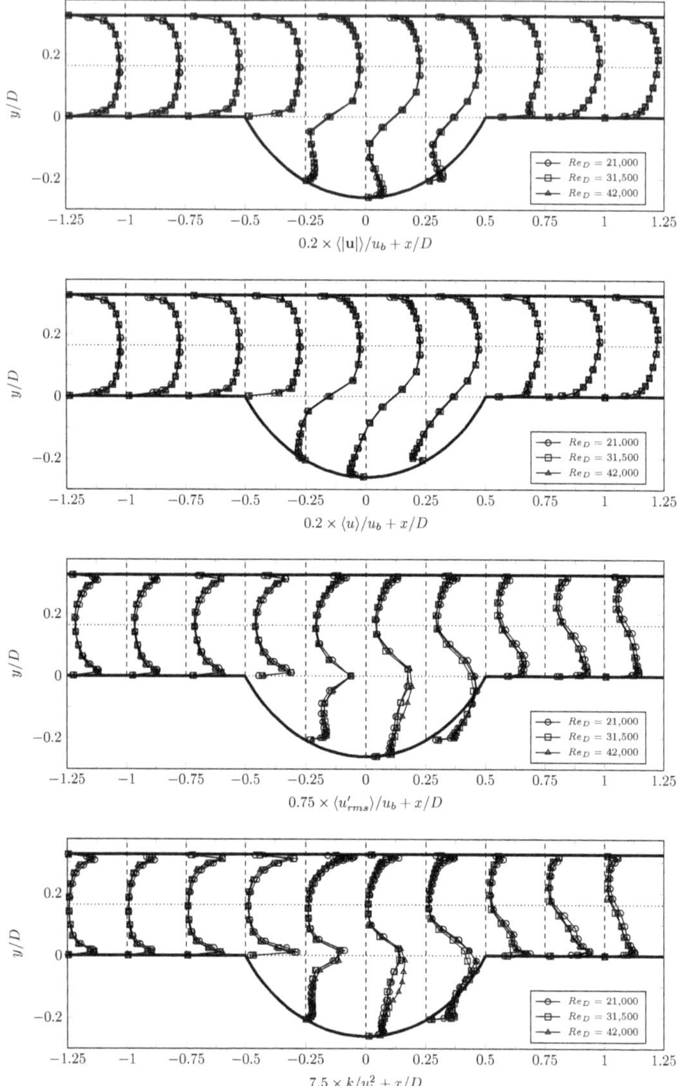

Fig. A.3: Influence of the Reynolds number Re_D on the unladen turbulent channel flow over a single spherical dimple with a dimple depth-to-dimple diameter ratio of $t/D = 0.26$.

A.2.4 Influence of the Reynolds number ($t/D = 0.35$)

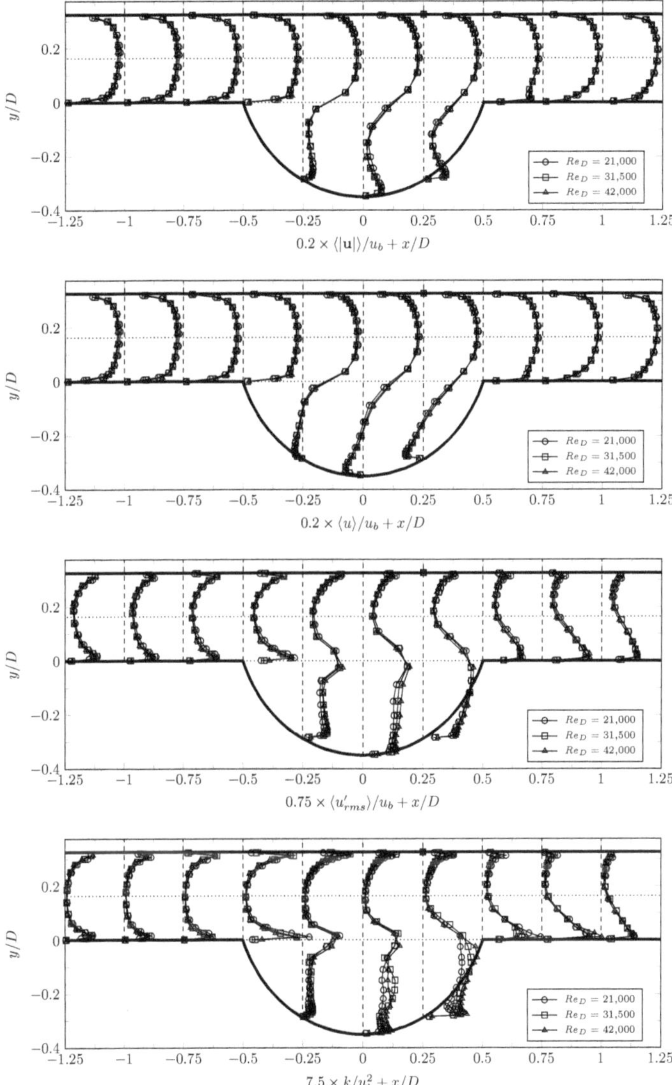

Fig. A.4: Influence of the Reynolds number Re_D on the unladen turbulent channel flow over a single spherical dimple with a dimple depth-to-dimple diameter ratio of $t/D = 0.35$.

A.2.5 Streamwise velocity distributions $\langle u \rangle / u_b$

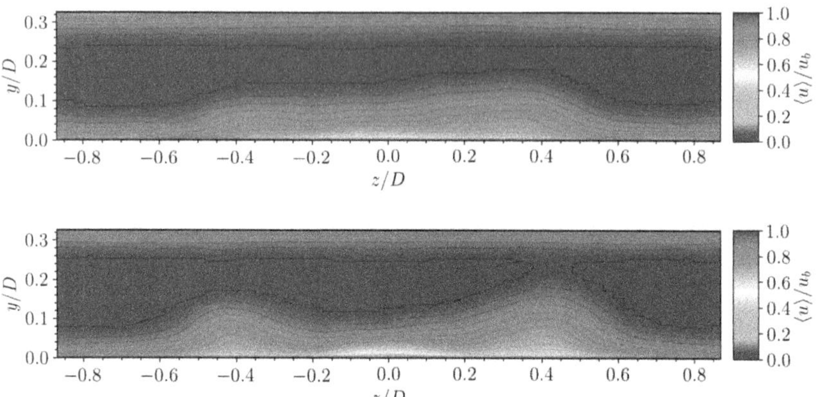

Fig. A.5: Distribution of the time-averaged streamwise velocity $\langle u \rangle / u_b$ in the $y - z$ plane at $x/D = 0.75$ for $Re_D = 42{,}000$ and a dimple depth-to-dimple diameter ratio of $t/D = 0.26$ (top) and $t/D = 0.35$ (bottom).

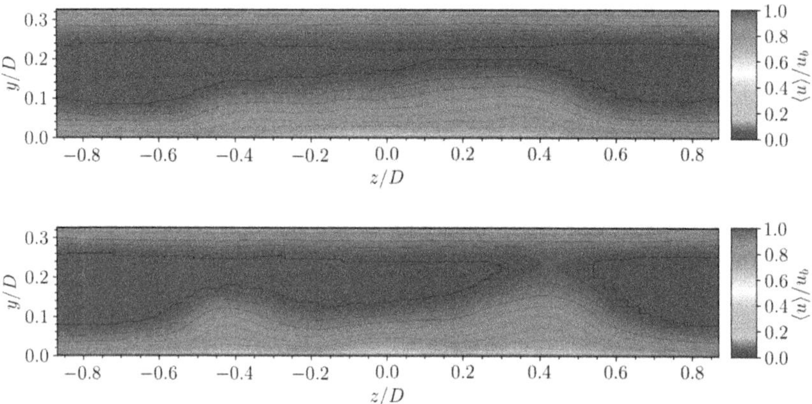

Fig. A.6: Distribution of the time-averaged streamwise velocity $\langle u \rangle / u_b$ in the $y - z$ plane at $x/D = 1.0$ for $Re_D = 42{,}000$ and a dimple depth-to-dimple diameter ratio of $t/D = 0.26$ (top) and $t/D = 0.35$ (bottom).

A.2.6 Vorticity distributions $\langle|\boldsymbol{\omega}|\rangle$

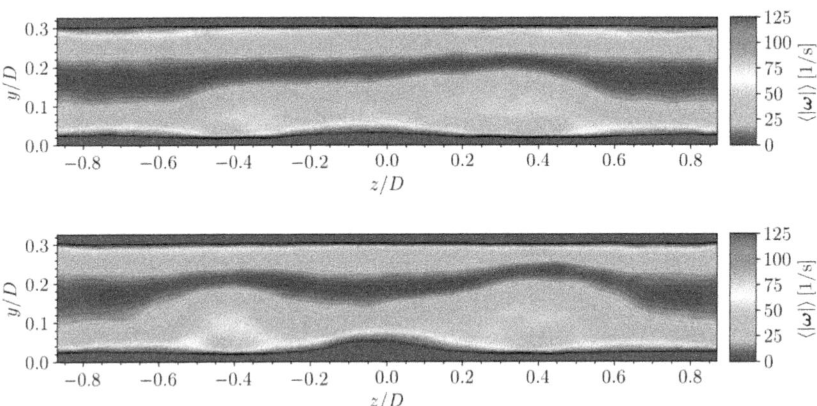

Fig. A.7: Distribution of the time-averaged vorticity magnitude $\langle|\boldsymbol{\omega}|\rangle$ in the $y-z$ plane at $x/D = 0.75$ for $Re_D = 42{,}000$ and a dimple depth-to-dimple diameter ratio of $t/D = 0.26$ (top) and $t/D = 0.35$ (bottom).

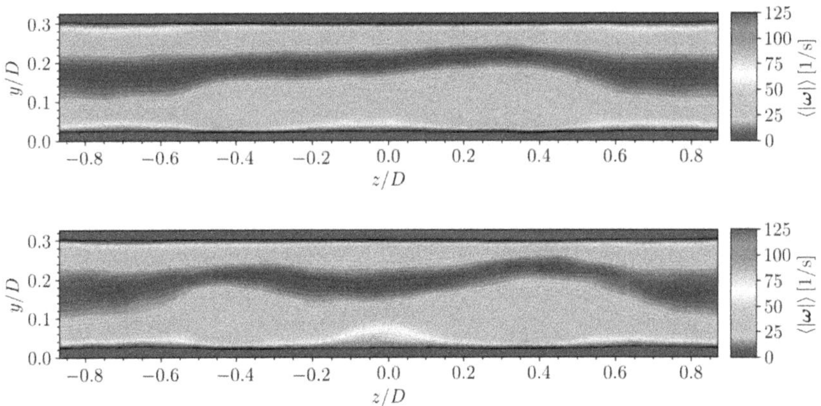

Fig. A.8: Distribution of the time-averaged vorticity magnitude $\langle|\boldsymbol{\omega}|\rangle$ in the $y-z$ plane at $x/D = 1.0$ for $Re_D = 42{,}000$ and a dimple depth-to-dimple diameter ratio of $t/D = 0.26$ (top) and $t/D = 0.35$ (bottom).

A.2.7 Vorticity distributions $\langle \omega_x \rangle$

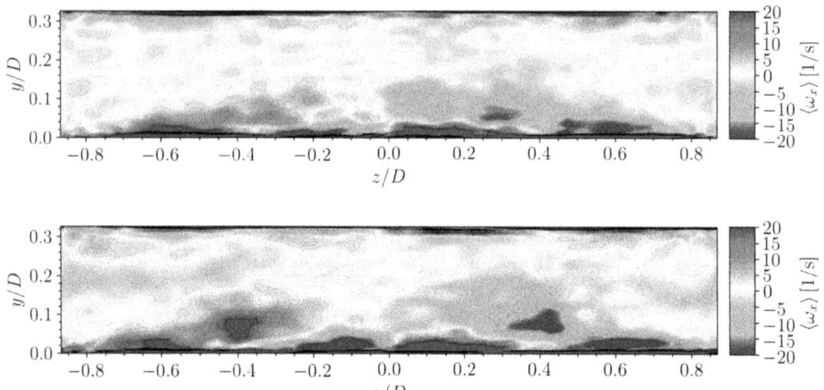

Fig. A.9: Distribution of the time-averaged streamwise vorticity $\langle \omega_x \rangle$ in the $y-z$ plane at $x/D = 0.75$ for $Re_D = 42{,}000$ and a dimple depth-to-dimple diameter ratio of $t/D = 0.26$ (top) and $t/D = 0.35$ (bottom).

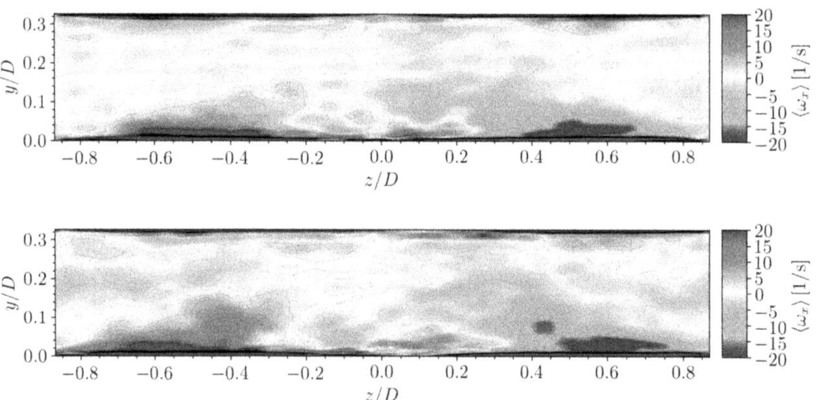

Fig. A.10: Distribution of the time-averaged streamwise vorticity $\langle \omega_x \rangle$ in the $y-z$ plane at $x/D = 1.0$ for $Re_D = 42{,}000$ and a dimple depth-to-dimple diameter ratio of $t/D = 0.26$ (top) and $t/D = 0.35$ (bottom).

Statutory declaration in lieu of an oath

I hereby declare in lieu of an oath that I have completed the present dissertation entitled *"Numerical modeling and simulation of particulate fouling on structured heat transfer surfaces using multiphase Eulerian-Lagrangian LES"* independently and without illegitimate assistance from third parties. I have use no other than the specified sources and aids. In case that the thesis is additionally submitted in an electronic format, I declare that the written and electronic versions are fully identical. The dissertation has not been submitted to any examination body in this, or similar, form.

Rostock, September 1, 2020 Robert Kasper

List of publications

Book contributions

Kasper R., Turnow J., Kornev N. (2019) Simulation of Particulate Fouling and its Influence on Friction Loss and Heat Transfer on Structured Surfaces using Phase-Changing Mechanism. In: Nóbrega J., Jasak H. (eds) OpenFOAM®. Springer, Cham.

Journal articles

Kasper R., Turnow J., Kornev N. (2017). Numerical modeling and simulation of particulate fouling of structured heat transfer surfaces using a multiphase Euler-Lagrange approach. Int. J. Heat Mass Transfer, 115, 932–945.

Kasper R., Deponte H., Michel A., Augustin W., Turnow J., Scholl S., Kornev N. (2018). Numerical investigation of the interaction between local flow structures and particulate fouling on structured heat transfer surfaces. Int. J. Heat Fluid Flow, 71, 68-79.

Turnow J., **Kasper R.**, Kornev N. (2018). Flow structures and heat transfer over a single dimple using hybrid URANS-LES methods. Computers & Fluids, 172, 720-727.

Kasper R., Turnow J., Kornev N. (2019). Multiphase Eulerian-Lagrangian LES of particulate fouling on structured heat transfer surfaces. Int. J. Heat Fluid Flow, 79, 108462.

Deponte H., **Kasper R.**, Schulte S., Augustin W., Turnow J., Kornev N., Scholl S. (2020). Local and time resolved investigation of particle deposition on dimpled heat transfer surfaces. Chem. Eng. Sci., 227, 115840.

Oral presentations / conference proceedings

Kasper R., Turnow J., Klunker J., Kornev N. (2015). Simulation of particle fouling and its influence on friction loss and heat transfer on structured surfaces using Phase changing mechanism. 8th International Symposium on Turbulence, Heat and Mass Transfer (THMT-15), September 15-18, 2015, Sarajevo, Bosnia and Herzegovina.

Kasper R., Turnow J., Kornev N. (2016). Numerische Simulation von Partikelfouling auf strukturierten, wärmeübertragenden Oberflächen. Jahrestreffen der ProcessNet-Fachgruppe Wärme- und Stoffübertragung, DECHEMA, Kassel, Deutschland

Kasper R., Turnow J., Kornev N. (2016). Simulation of Particle Fouling and its Influence on Friction Loss and Heat Transfer on Structured Surfaces using Phase Changing Mechanism. 11th OpenFOAM Workshop (OFW11), 26-30, June 2016, Guimarães, Portugal.

Kasper R., Turnow J., Kornev N. (2016). Simulation of Particle Fouling and its Influence on Friction Loss and Heat Transfer on Structured Surfaces using Phase Changing Mechanism. 7th International Conference on Vortex Flows and Vortex Models (ICVFM7), September 19-22, 2016, Rostock, Germany.

Kasper R., Turnow J., Kornev N. (2017). Thermo-hydraulic analysis of structured heat transfer surfaces under consideration of particulate fouling using a multiphase Eulerian-Lagrangian method. Heat Exchanger Fouling and Cleaning Conference XII, June 11-16, 2017, Aranjuez (Madrid), Spain.

Turnow J., **Kasper R.**, Kornev N. (2017). Flow structures and heat transfer over a single spherical dimple using hybrid URANS-LES methods. 10th International Symposium on Turbulence and Shear Flow Phenomena (TSFP10), July 6-9, 2017, Chicago-IL, USA.

Kasper R., Turnow J., Kornev N. (2017). Numerical investigation of the interaction between local flow structures and particulate fouling on structured heat transfer surfaces. 10th International Symposium on Turbulence and Shear Flow Phenomena (TSFP10), July 6-9, 2017, Chicago-IL, USA.

Kasper R., Turnow J., Kornev N. (2017). Analysis of the thermo-hydraulic performance of structured heat transfer surfaces under consideration of particulate fouling using a multiphase Euler-Lagrange approach. 12th OpenFOAM Workshop (OFW12), July 24-27, 2017, Exeter, UK.

Kasper R., Turnow J., Kornev N. (2018). Eulerian-Lagrangian LES of particulate fouling on structured heat transfer surfaces. 9th International Symposium on Turbulence, Heat and Mass transfer (THMT-18), July 10-13, 2018, Rio de Janeiro, Brazil.

Kasper R., Turnow J., Kornev N. (2019). Prediction of particulate fouling on structured surfaces using multiphase Eulerian-Lagrangian LES. 11th International Symposium on Turbulence and Shear Flow Phenomena (TSFP11), July 30 - August 2, 2019, Southampton, UK.

Posters

Klunker J., Turnow J., Kornev N., **Kasper R.** (2015). Wärmeübertragung auf strukturierten Oberflächen unter Berücksichtigung von Partikel-Fouling. Jahrestreffen der ProcessNet-Fachgruppe Wärme- und Stoffübertragung, DECHEMA, Leipzig, Deutschland.

Kasper R., Turnow J., Kornev N. (2017). Thermo-hydraulische Analyse strukturierter Oberflächen unter Berücksichtigung von partikulärem Fouling. Jahrestreffen der ProcessNet-Fachgruppe Wärme- und Stoffübertragung, DECHEMA, Bruchsal, Deutschland.

Deponte H., **Kasper R.**, Michel A., Turnow J., Augustin W., Kornev N., Scholl S. (2018). Experimentelle und numerische Untersuchung des Ablagerungsverhaltens von Partikeln an Dellenstrukturen. Jahrestreffen der ProcessNet-Fachgruppe Wärme- und Stoffübertragung, DECHEMA, Bremen, Deutschland.

Lebenslauf

Name: Paul Robert Kasper

Geburtstag: 05. April 1988

Geburtsort: Magdeburg

Schulausbildung:

07/2000 - 07/2007 Kurfürst-Joachim-Friedrich-Gymnasium, Wolmirstedt.

Zivildienst:

10/2007 - 06/2008 Klinik für Neurochirurgie, Universitätsklinikum Magdeburg A. ö. R., Magdeburg.

Hochschulausbildung:

10/2008 - 10/2012 Bachelorstudium (B.Sc.) Maschinenbau, Otto-von-Guericke-Universität Magdeburg.

10/2012 - 11/2014 Masterstudium (M.Sc.) Schiffs- und Meerestechnik, Universität Rostock.

Beruflicher Werdegang:

seit 01/2015 Wissenschaftlicher Mitarbeiter, Lehrstuhl für Modellierung und Simulation, Universität Rostock.

Forschungsschwerpunkt: Numerische Modellierung und Simulation von partikulärem Fouling auf strukurierten, wärmeübertragenden Oberflächen.

www.ingramcontent.com/pod-product-compliance
Ingram Content Group UK Ltd.
Pitfield, Milton Keynes, MK11 3LW, UK
UKHW022001190726
13853UKWH00004B/1657